Environment and Mental Health

Stephen M. Williams

JOHN WILEY & SONS
Chichester · New York · Brisbane · Toronto · Singapore

Published in 1994 by John Wiley & Sons Ltd,
Baffins Lane, Chichester,
West Sussex PO19 1UD, England
Telephone National Chichester (0243) 779777
International +44 243 779777

Other Wiley Editorial Offices

John Wiley & Sons, Inc., 605 Third Avenue,
New York, NY 10158-0012, USA

Jacaranda Wiley Ltd, 33 Park Road, Milton,
Queensland 4064, Australia

John Wiley & Sons (Canada) Ltd, 22 Worcester Road,
Rexdale, Ontario M9W 1L1, Canada

John Wiley & Sons (SEA) Pte Ltd, 37 Jalan Pemimpin #05-04,
Block B, Union Industrial Building, Singapore 2057

Library of Congress Cataloging-in-Publication Data

Williams, Stephen M.
Environment and mental health / Stephen M. Williams.
p. cm.
Includes bibliographical references and index.
ISBN 0-471-95002-5 — ISBN 0-471-95093-9 (pbk.)
1. Mental illness — Environmental aspects. 2. Mental health — Environmental aspects. 3. Environmental psychology. 4. Mental illness — Etiology. I. Title.
RC455.4.E58W55 1994
616.89 — dc20 94-1707
CIP

British Library Cataloguing in Publication Data

A catalogue record for this book is available from the British Library

ISBN 0-471-95002-5 (cloth)

Typeset in 10/12pt Palacio from author's disks by
Mathematical Composition Setters Ltd, Salisbury, Wiltshire
Printed and bound in Great Britain by Biddles Ltd, Guildford, Surrey

To my wife Brigitte,

Syn Athene kai cheiron seauton kinei
With Athena let your hand move too

my mother, Betty Hubbard,

I have a lot to thank you for!

and Roy Hubbard,

the father who was there

I have seen something else under the sun

The race is not to the swift
or the battle to the strong
nor does food come to the wise
or wealth to the brilliant
or favour to the learned
but time and chance happen to them all

Ecclesiastes 9:11

"How do you do?" Miss Barkley said. "You're not an Italian, are you?"

"Oh, no."

Rinaldi was talking with the other nurse. They were laughing.

"What an odd thing—to be in the Italian army."

"It's not really the army. It's only the ambulance."

"It's very odd though. Why did you do it?"

"I don't know," I said. "There isn't always an explanation for everything."

"Oh, isn't there? I was brought up to think there was."

E. Hemingway, *A Farewell to Arms*.

Contents

Figures

Preface

What does the environment have to do with mental health? Most people can see that there is an issue whether we are born or made mentally healthy. Since we are made what we are by environments, a weak connection starts to appear. Yet surely the primary association of "environment" nowadays is to conservation of the physical environment, to "greening"?

That is true, but it is an association in people's minds, which develop new links constantly. Is a green environment really so remote from mental health? Is not the idea of escape from our cherished yet constraining civilisation part of the hope of self-renewal? And how about the inspiration to reach out to new worlds in space?

It is already time to come a bit clean. This book began while I served my notice period before leaving the University of Ulster, where I had taught a physically oriented course on Environmental Psychology for several years. It is ending as I work at the Institute for Health Studies in North Essex, still a psychologist, concerned, inevitably, with mental health. I am trying to ford a river crossing my own life, and whether it will work for you depends, I suppose, on how alike we are.

The Ulster course originated in an odd way. During a short period as a research consultant at the Open University's Institute for Education Technology, I put together the plan of a new sort of Psychology course that would cover the interests left undimmed by the 6-year grind of a D.Phil. in Experimental Psychology. Casting around for the most appropriate label, "Environmental Psychology" was what I lit upon, and I even submitted an unsuccessful grant proposal for help with teaching innovation. At Ulster, lo and behold, I found that my predecessor as Social Psychologist, G.W. ("Bill") Mercer, who had returned to Canada and whom I have unfortunately never met, had introduced a course in Environmental Psychology. It already existed. It was healthy in North America, complete with textbooks, and I took my main task as being to establish it at my new University.

Chapter 1 (as well as 2 and 3) is heavily influenced by my time at Ulster. A version of this was in the draft seen by Belhaven's very helpful readers, who themselves suggested the orientation towards mental health that is taken up by Chapters 4, 5 and 6. What both of them picked out about it was my attitude to religion *vis-à-vis* science. I hope the chapter now makes this clearer, but, timorously, something about myself may be helpful here. I say timorously, because I agree with Brian Appleyard (in *Understanding the Present*), that there is a big element of "is there an easy way to ignore this nut?" in a lot of people's questioning nowadays of others about religion. What I am happy to disclose is that for long periods of my life I have been a regular churchgoer. I believe there is a God, and tend to put down most people's doubts about it either to psychologically insensitive religious education or to a reaction against religious fundamentalism. None the less, I have struggled with myself since childhood over the idea of a personal God, which most would regard as basic to Christianity, and though I feel much closer to the idea now, that struggle continues. The Bible has been a lifelong source of strength and comfort to me, and has only, I feel, to be read in the right way.

Behaviourist environmentalism is distasteful to many because of its determinism (though geneticism is surely no less so). I was taught by Cambridge philosophers that determinism and free will are compatible, and as a consequence find it difficult to be bothered about whether determinism is true.

Much early Psychology sought to establish causal influence of "independent variables" on "dependent variables". Since an *interesting* causality is often an illusion, there has been trouble enough over this. When psychologists show, in an experiment that aspires to be scientific, that a limited short-term memory store causes forgetting, the public greet their efforts with a resounding 'so what?' Noticing our own forgetting may lead us to try to do something about our short-term store, with mnemonics, or even pharmacology. This is what is meant by "interactionist" thinking, which has been the modern trend in Environmental Psychology. Still deterministic though.

Acknowledgements

Intellectual debts may be more important than practical ones, though I doubt it. In any case, I owe so many. To mention is to omit, and so to offend. Better just to keep quiet. But justice compels me to acknowledge a debt to Terence Lee, the doyen of British Environmental Psychologists, who read for John Wiley & Sons, and made, so well, so many thought-provoking comments.

The book would never have been written without some remarkable flexibility shown by my employers, the Institute for Health Studies in Colchester. It is something off the beaten track so far for Health Service educational establishments, and to stray from tradition takes daring. In the flesh, this means acknowledging the indispensable support and forbearance of David Braithwaite, Director of the Institute, and Mohammad Abuel-Ealeh, my Head of Department. I should also like to thank the Director of the Common Foundation Programme of the Nursing Registration Diploma, Hilary Bebb, who has been able to use less teaching from me than she otherwise could.

Others at the Institute I should like to thank include Martin Hillary and George Waterman, for helpful discussion about psychological issues. Many of my ideas about mental health have been formed by members of the Mental Health Team led by Paul O'Loughlin. Bernard Anderson gave much computer support, Edward Smith helped with illustrations, and I should mention Shirley Saunders, Lindsey Williams and Pam Sheehan of the library. This is not to forget staff of the Albert Sloman Library of the University of Essex, and the Coleraine Library of the University of Ulster. Stephen Power took the photographs for Figures 2.5, 5.1 and 5.2.

My wife Brigitte discusses my ideas with me, as well as justifying to me each day the best decision I ever made.

Iain Stevenson of John Wiley & Sons is a publisher out of the ordinary. He has gone to great pains with me in our negotiations, given me a huge amount of help with the form of the book, and even added expert comments on the first draft. Through two years of collaboration over this, he has never failed to answer a query or sustain my energy. My thanks are also due to the John Wiley production team.

None of these bears any responsibility for flaws in the book, which are mine alone.

CHAPTER 1

Introductory: Why Scientific Psychology has Neglected Environment

Though your beginning was small, yet your latter end should greatly increase.
Job 8:7

PSYCHOLOGY AS SCIENCE

To recognise the consequence of the concept of "environment" in Psychology we need first to have some conception of the nature of Psychology as an intellectual project. My main thesis is that Psychology has neglected environment as an explanation of behaviour. What could it be about Psychology that would cause this? Since the founding of Wundt's laboratory in the 1870s, the received or canonical view of Psychology has been that it is a branch of science. Because science finds the real environment too difficult and complex to master, it is ignored. As we shall see, when it comes to mental health, this leads to pernicious outcomes.

Science is the great cultural force of our time; it touches every branch of study. When the study of language can be called "linguistic science" and the study of politics "political science", it is not surprising that the study of mind too should be called a "science".

I argued in my previous book *Psychology on the Couch* that psychology does not in fact measure up to a description as "science". But I take the description more seriously than does Heather, who comments acerbically (1976: 10): "To call this collection [academic Psychology] of irrelevance, triviality and downright silliness a science is preposterous." This raises further questions as to whether Psychology *can* be or *should* be a science, since Heather goes on (p12) to say that psychologists use science to dehumanise man.

Within the discipline there is now a branch known as "Environmental Psychology". Will this be able to sidestep the fact that Psychology is generally treated as a science? We need a better understanding of the scientific approach so as to comprehend the actual as well as the potential meaning and significance of "Environmental Psychology" within it.

SOCIETY SUPPORTS SCIENCE

People often think of science in the abstract, as the hotline to truth, independent of human frailty. On the contrary, science is made by scientists. Scientists and therefore science are supported by the wider society and have become so supported as part of a historical development. John Roberts (1985) has said: "The crucial change in the making of the modern mind was the widespread assumption of the idea that the world is essentially rational and explicable, though very wonderful and complicated". It was this assumption that made public support of science possible, with the scope of its domain ever broadening to include ambitious projects like Psychology—the explanation of mind itself.

Such projects are flawed, according to some critics, by the flawed nature of society itself. Thus for Marx:

> [under capitalism] nature becomes purely an object for mankind, purely a matter of utility; ceases to be recognised as a power for itself, and the theoretical discovery of extraneous laws appears merely as a ruse so as to subjugate it under human needs, whether as an object of consumption or as a means of production [*Das Kapital*].

None the less capitalism is the actual social organisation of most of the world; Marx himself would acknowledge the usefulness of science to capitalism, and under capitalism science has thrived. In times of general war, too, the public revalues not only soldiers but also scientists. If science serves capitalism and the war machine, it is unlikely to undermine these by showing how inequalities come from people's environments.

Sociologists tend to argue from such phenomena as "multiple discovery" —simultaneous invention by independent scientists—that not only time of war but other general factors in the social context ease or hinder creative developments. But it has often been claimed that the creative developments of science produce a special kind of knowledge. All the major contributors in the nineteenth century to the "sociology of knowledge" in its beginnings were doubtful about the possibility of including natural science within its scope (Mulkay, 1979: 3). So it is possible to argue that scientific knowledge depends much less than other kinds of knowledge upon the society within which it is created. There is probably a lot to this with regard to the sciences of matter, but for Psychology understanding its social basis will prove illuminating. Scientists are regarded as a modern priesthood, with privileged access to truth. That is why society supports them. It is natural enough that many psychologists wish to join that priesthood, if they can carry it off.

NATIONS AND SCIENCE

One way to understand the social basis is to look at the nation producing the science, indeed "nation" seems more concrete as an idea than "society". There are obvious national differences in science production: science is above all the product of the West.

Perhaps I speak too soon? The Frenchman Bailly (see Jaki, 1978), writing while his country led the world in science, asserted the cradle of science was Mongolia. This is a quaint example of an "origin myth", found even in something so avowedly rational as science. Supposed revolutions are a favourite origin myth for new fields of enquiry, such as, today, "Cognitive Science".

Nations vary in their trends of science production. The Soviet Union contributed less than 1% of the world's total scientific papers in 1910, but 18% by 1960. During the same period Germany saw its share fall from 40 to 10%, while the British Commonwealth countries contributed a constant 15%. One could speculate endlessly about the reasons for these differences, but they underline the dependence of science upon its social context.

An interesting example of the inter-relationship of the general culture and science is Spain, once "Top Nation" (says *1066 and All That*). Spanish culture has been deficient in science, and this has contributed to material backwardness (though this is rapidly changing) of the country. Unamuno said the living belief in the immortality of the soul has given Spain a tragic sense of life that makes material things seem unimportant and hinders the motivation towards disinterested scientific discovery. Another theory would be that it was before the scientific age that her huge empire was conquered, and this bred a complacency and indifference to the Industrial Revolution.

The great engine-house of modern Psychology has been America. After Wundt founded his laboratory at Leipzig, Americans were finding that the Doctor of Philosophy degree, also a German invention, was the path to an academic career. Many went to gain it under Wundt himself in this morning of Psychology, but were deadened and depressed by working with him (Leahey, 1987: 200). The optimism of these pioneers, often radical environmentalists when they arrived, was suppressed. Geneticist perspectives, such as that of phrenology (which reads character from cranial bumps), were strong in America. Young (1970) in his book *Mind, Brain and Adaptation in the Nineteenth Century* focussed on the work of Europeans like Gall and Spurzheim. This needs to be supplemented by consideration of the American arm of phrenology, which was largely Orson and Lorenzo Fowler. These brothers minimised the scientific content of phrenology and maximised the practical applications. They set up an office in New York where clients could have their character read for a fee. They wrote endlessly of the benefits of phrenology and published a journal that endured from the 1840s to 1911 (Leahey, 1987: 247). It would be wrong to infer from the success of behaviourism that early American Psychology was predominantly environmentalist.

Discussion of national differences in science production should not leave the impression that modern science is wholly in the hands of the North. Less developed nations are tragically aware (Kipling) that "Whatever happens we have got/The Maxim gun and they have not". They have themselves turned to science "and to optimism, rationalism and individualism—the ideological Trinity of the civilisation from which modern technology emerged" (Roberts, 1985).

RELIGION AND SCIENCE

The historical development of science cannot be understood except in the context of the traditional Trinity, of Christianity, once the framework for all intellectual endeavour in the West. For Condorcet the two meant Galileo and the rack, irreconcilably opposed (Fig. 1.1). Like most simple pictures, Condorcet's has come to seem less compelling (Jaki, 1978: 1).

Figure 1.1 Galileo recants. Reproduced by permission of the Mansell Collection

Condorcet said the Catholic church preferred Gospel to science; I should say this must be less true today than in the mid-nineteenth century, and the preference is not blind. There are different strands within the Church. As early as the debate over evolution, evangelicals, as opposed to fundamentalists, recognised the possible parallel between Darwin and Galileo, and the dangers of over-literal readings of the Bible. None the less the fact should be recognised that early science did have a continual struggle with the Church. One effect of that, I would contend, has been a defensive stance of science. It has concentrated upon the simple rather than anything so complex as the environment, reluctant to venture into broad questions of life that the Church saw as its own preserve.

Yet almost as loud as the voices proclaiming science as a replacement for Christianity are those saying the two are compatible and intertwined. Prehistoric man probably believed time revolves in eternal cycles. Christianity identified a moment of origin and a historical sequence leading to the Incarnation in Jesus. It has often been said that early science could not have begun without this changed conception of time as linear rather than cyclical, and of *progress* along that time-line. Isaac Newton, whom many see as the true founder of modern science, wrote in a letter a significant declaration.

> When I wrote [*Principia*] I had an eye upon such principles as might work with considering men for the belief of a Deity and nothing can rejoice me more than to find it useful for that purpose [Westfall, 1981].

This was certainly typical of centuries earlier than our own. A noted early biologist Swammerdam opened a lecture, without irony: "Here I bring you proof of God's providence in the anatomy of a louse" (Lindeboom, 1975). He was voicing the common assertion that the world is revealed by science to have been designed by an intelligence. But most philosophers today agree that this traditional theological Argument from Design was destroyed by David Hume.

With regard to scientific Psychology specifically, John Broadus Watson, like the majority of early behaviourists, first chose the ministry as his vocation, before turning to the field that made him famous. It is as though at this time a thirst for the fundamentals of life, that had once been slaked by religion, was now better served by Psychology.

It could be argued that part of the Christian message is acceptance of life as it is, of the fact of personal death, and submission to God and fate. That was what Marx disliked about it, with his description of it as "the opium of the people" and his "philosophers have sought to interpret the world, the point is to change it". Marx saw himself as a scientist, and if there is a "message of science" it is the call to mastery of nature and challenge of the old chains.

NONCONFORMISM

I contended that science was forced by its early relations with the Church to adopt a defensive and over-simplified approach. The early Protestant church was by its nature less restrictive than the Catholic church it broke away from. Max Weber (1946) said it was Protestantism and particularly the nonconformist branch of Protestantism that were associated with the rise of science. This was a major theme of the historical sociology of R.K. Merton (1965), who documented, for example, the way Puritan attitudes were so influential in the early days of the Royal Society. In the (science-based) Industrial Revolution, there is evidence that early English entrepreneurs were predominantly Nonconformists. In the broader sense too, nonconformism seems associated with the scientific spirit. Thus Joseph Needham (1954), the historian of Eastern science, says that in the past the great bureaucracies (e.g., Persia and China) were everywhere hostile to independent scientific thought.

International sociological comparisons also suggest that where the national organisation of science is decentralised and encourages competition (i.e. is not a monolithic home for "organisation man") productivity is greatest (Ben-David, 1962). Though, today, nation-states are overwhelmingly the predominant financial supporters of science, there is a conflict between the scientist's need for autonomy and a government's need for security and secrecy. This conflict showed itself very starkly in the McCarthy era in the USA.

There is an authoritarian trait in human nature that holds back science. There is abundant evidence in the history of science that when discoveries are made by scientists of lower standing they may be resisted. Scientists of higher standing use detrimentally the authority that the higher position provides. A classic illustration of this comes from the career of Rayleigh, the great nineteenth-century physicist. One of his papers was accepted by the British Association for the Advancement of Science only after his name, accidentally omitted in the first submission of the paper, was supplied. Creativity is needed in science, and this may be allied to emotional and irrational behaviour that is far from a hallmark of conformity. Issues of conformity and consensus are also raised by Kuhn's (1962) idea of the existence of agreed "paradigms" as what makes a discipline "science". Before there was a philosophy of science, people assumed that scientists agree because there is an objective truth for them to agree about. But scientists are people too, and people agree for many reasons: to avoid treading on toes, to keep in with the boss, to make friends. Kuhn introduced the idea of revolutionary breakdowns of scientific paradigms, that has been well supported with examples by Bernard Cohen's *Revolution in Science* (1985).

THE CONCEPT OF PROGRESS

This nonconformism goes with a positive attitude towards *progress*. This attitude goes back to the beginning of the scientific revolution. Take the words of one of the seminal thinkers, Francis Bacon, Lord Verulam. "Let man endeavour an endless progress or proficience in both the book of God's word and the book of God's work" (1620). Conformity to tradition retards progress. Progress is an essential feature of science: it is said that science moves on while other disciplines simply move about. Thus,

> The work of art of a period that has worked out new technical means, or, for example, the laws of perspective, does not stand "higher" . . . A work of art which is genuine "fulfilment" is never surpassed; it will never be antiquated [Weber, 1946: 137].

Progress in science often goes with progress in society; since the scientific revolution writers have made progress into a totem, in Herbert Spencer's words: "not an accident but a necessity" (1876—1896). A thorough historical perspective would show, of course, that economic progress did not begin with the scientific revolution. For example, productivity in British agriculture roughly doubled between 1400 and 1600 without any new inventions at all.

Some science, on the other hand, has an internal dynamic divorced from wider society: Oppenheimer, one leader of the atom bomb project, said that scientists are actually ready to tackle any problem that is "technically sweet".

The link between scientific and social progress can be seen clearly in the nineteenth-century debate over evolution. The doctrine of evolution was resisted fiercely by political conservatives, who saw an analogy between the fixity of species and the fixity of social structural arrangements, i.e. social stability. It is probably no accident that in France, where people were mindful of the recent cataclysmic Revolution and Terror, Darwinism made heavy going.

Creative writers, who I think may mirror their society better than anybody else, vary markedly in their attitudes *pro* or *contra* science and progress. Defoe's *Robinson Crusoe* is in sharp contrast to Golding's *Lord of the Flies*, and W.H. Auden's positive attitude the complete opposite of T.S. Eliot's.

Environmental Psychology is generally regarded as a branch of Social Psychology. If Social Psychology is science, it should be making progress. Yet it has been persuasively argued that it is not: "Knowledge [about human interaction] cannot accumulate in the usual scientific sense because such knowledge does not generally transcend its historical boundaries" (Gergen, 1973). Gergen's view has been strongly criticised: "This is analogous to claiming that no universal theories could have been developed in the natural sciences because ice changes into water ... or dinosaurs are no longer with us" (Schlenker, 1974). As examples of social theories that *are* transhistorically valid Schlenker gives those known as social learning, social facilitation, social comparison and mere exposure. He also argues that if social processes were as transient as Gergen believes, it would be difficult to explain why the writings of Aristotle, Plato, Rousseau, Hobbes, etc. are still used. The Gergen–Schlenker debate has drawn in further participants and the interested reader can of course look at the original articles. My own view is that Gergen has indicated an important difference between Social Psychology and traditional forms of scientific inquiry. It is because they are so difficult to treat scientifically, that the social in general, and the environmental in particular, have been so neglected in psychological study. The baby has been thrown out with the bath water.

What goes wrong with "stride of progress" ideologies is what is often called "Whiggism", the tendency to see the past as merely the inevitable precursor to, and handmaiden of, the present. Jaki (1978: 36) says that the essays of Turgot, the last great statesman of the French *ancien régime*, "were unwitting proof that any age which looks upon itself as the epitome of this or that perfection inevitably loses its sense of history, which is always a march and never an arrival". This error, so obviously blind in Turgot's case, has been repeated in every age, so that critics of it, like Gergen, are still needed.

Hegel's idea about historical change was that, rather than simple progress, it was a "dialectic", in which people reacted against the past, and then found a reconciliation ("synthesis") of the reaction and of what was reacted against. Hegel was interested in the historical development of ideas, but Marx adapted the dialectic to understand changes in material production (economics), i.e., "dialectical materialism".

Hegel's analysis is easy to apply to Psychology. A historian of Psychology, Leahey (1987) writes that in 1920s America,

> Psychology was king ... One had only to read the newspapers to be told with complete assurance that psychology [then meaning environmental factors rather than original sin] held the key to the problems of waywardness, divorce and crime ... By 1930 the fad for psychology had run its course. After the crash of 1929, the popular press had more pressing economic matters to consider, and the volume of pieces written on psychology diminished noticeably.

Since when Psychology has certainly had new vogues and out-periods.

I have talked, too, about the philosophy of science. Here, as well, it has been claimed that Feyerabend, Kuhn and Lakatos were reacting against the work of Karl Popper, just as the latter was rejecting inductivism and the verifiability principle of the Vienna circle (Richards, 1983: 60).

CONTROVERSY IN SCIENCE

The simple view of science as objective truth encounters a problem with the prevalence of controversy in science and scientific Psychology. It seems evident that engaging in controversy is not engaging in progress, yet Kuhn's theory of the history of science gives controversy during periods of "revolutionary science" a central role. During such periods,

> It is not of course that logical argument has no part to play, but rather that the revolutionary crisis airs a variety of fundamental standards and metaphysical assumptions that can be resolved—at the time of the crisis at any rate—only in terms of intuitive judgment [Richards, 1983: 65].

So Kuhn's picture attacks the centrality of rationality, let alone progress, in science. The power of irrational and social factors is also captured by Max Planck's much-quoted, but hopefully somewhat pessimistic, claim, "A new scientific truth is not usually presented in a way that convinces its opponents; rather they gradually die off, and a rising generation is familiarised with the truth from the start" (Heilbron, 1986).

Psychologists are unsure of themselves as scientists so, to maintain appearances, controversy is anathema: "We attribute controversy to the Age of Schools and we make it altogether clear that we are well past that age" (Henle, 1986: 93). But human nature will out. There was a virulent debate earlier in the century over whether "imageless thought" is possible (a debate that may yet come back to haunt us). Among early American psychologists Titchener, an immigrant from England, was probably foremost. It was important, he thought, to deny categorically the claim of Würzburg introspectionists to think without images. Even for such a difficult case as the concept of "meaning" he described his own vivid images as follows: "I see meaning as the blue-grey tip of a kind of scoop, which has a bit of yellow above it . . . and which is just digging into a dark mass of . . . plastic material". Does this give the idea? Yet this acrid and verbose controversy has been called "a Kuhnian paradigm clash"—the effect of it was to aid and abet the birth of behaviourism. It is an illustration of why some believe time in the behavioural sciences is still cyclical rather than linear. By the second edition (1955) of Woodworth's & Schlosberg's *Experimental Psychology*, the controversy, which occupied many pages in the 1938 edition, had shrivelled to four paragraphs. Today it is rarely mentioned. But today cannot even keep its story straight. One examination of seven leading modern textbooks in Cognitive Psychology found of a total of 3246 publications only 19 (0.6%) were cited by every author; only 146 (4.5%) were cited by as many as three texts. There is simply no agreement yet on what is the progress which Psychology is supposed to have made. Pick your own objective truth.

EMPIRICISM AS A VALUE IN SCIENCE

A survey of popular opinion showed that for most people science is simply thorough and intensive study, and concluded: "very few really understand the scientific approach and about half say science can study *anything*." I prefer the popular usage, but scientists themselves certainly mean something else by "science". I stated in *Psychology on the Couch* my belief that science is defined best in terms of its values. These include system, rigour and precision, but neglect creativity, authenticity and richness, which might lead closer to environment. Another perspective comes from Sociology, which since R.K. Merton (1965) has accepted that science has certain institutional norms, usually labelled "universalism", "communism", "disinterestedness" and "organised skepticism". But this sits uneasily with the fact that much expenditure on science, particularly in the USA and UK, is for military purposes and does not even increase economic growth.

Scientists tend to be "doers", for example political activists, and something about their work makes them persevering, detractors would say "perseverative". It is widely agreed that a central cognitive value of science is empiricism – the confident appeal to experience as the touchstone of truth. This is part of the philosophy known as "positivism". It was for Descartes' emphasis on reason rather than experience that d'Alembert blamed him for the stagnation of French science in the following period. Experience is experience of an environment so the scientific approach might be thought sympathetic to environmentalism.

Yet it is easy to be misled by a canonical rather than veridical account here. Is science really so faithful to experience? It is my impression that many scientists are fascinated by the paranormal and by "science fiction". It is not my intention to knock these, and science fiction is often ahead of science. An early example is the specification in *Gulliver's Travels* of two satellites for Mars 150 years before their actual discovery. Less anecdotal is the modern recognition that: "Scientific knowledge . . . is not based upon experience as such, but upon the experience of the scientific profession" (Barnes, 1985: 45). That is, "experience" is an abstract idea until we make it concrete by specifying whose experience of what. Therefore many aspects, I would argue the most important aspects, of the experienced environment are neglected by science and by scientists. An historical illustration of this is the meteorite controversy. Many people saw and see meteors; few (and in the days of the controversy, mainly "common people") found meteorites on the land. Consequently science was painfully slow to draw a connection between the two. Barnes's most vivid example of the "closed circle" in science, drawn from an earlier time but merely throwing into relief tendencies still present today, is the procedure of the great seventeenth-century natural scientist Boyle. When important experiments were undertaken, Boyle would assemble witnesses and record their names and their qualifications, so as to give weight to their testimony and therefore to the standing of the experimental results reported. Witnesses had to be credible witnesses, for which it was requisite that they possessed sufficient social as well as scientific standing (Barnes, 1985: 51). The requirement of social standing is of course diluted in the world since the *ancien régime*, but not the idea of credible witness and the sociological factors that go with it. Positivism ignores this, and the death of positivism has been reported too often now to be readily believed—within Psychology an influential current philosophical school is the "unified positivism" of A.W. Staats (1991).

HANDLING COMPLEXITY

The great appeal of the scientific world-view is that through its theories it *simplifies* the complicated world around us.

Even at the outset of the scientific revolution ours was a bewildering world. It was a world in which "Red men scalped one another on the shores of the Great Lakes so that a Prussian king could keep a province stolen from his neighbour" (Macaulay, 1843). Any islands of clarity were welcome. Today we have moved into a world in which, Dubos (1965) claims, the tempo at which man changes the environment and his views of himself is now very rapid. It is so rapid that even the rules of right conduct must be changed from one generation to the next. Scientific simplification was welcome even though it was achieved at the price of what Weber (1946) christened *die Entzauberung der Welt*—the disenchantment of the world. The trouble was, claims Berman (1984) in his book *The Reenchantment of the World*, the increasing inability of the scientific world-view to explain the things that really matter. The complexity of the environment is simply too great to be amenable throughout to scientific methodology.

The line of response to this, if we are to retain something at least of the scientific approach, is to "loosen" science as much as possible, as Feyerabend (1975) does with his "epistemological anarchism" ("the only rule is: Anything goes"). Feyerabend could be called a maverick; but he himself claims for his own camp other more mainstream philosophers of science such as Lakatos. Lakatos held that science is a collection of "research programmes", some of which encounter "degenerative problem shifts" (i.e., turn into a modern version of scholasticism) and are discarded (Lakatos and Musgrave, 1970). Feyerabend likes Lakatos's account because, he claims, it is merely anarchism in disguise.

I should say that the response of Lakatos to complexity is, at least in part, that it should be laundered away. Thus he holds that the history of science ought primarily to be rational reconstruction. The historian should show only in footnotes how things really did go, criticising history for its deviation from the path of true reason. So the power of science to simplify must be exaggerated by the reconstructed accounts of it we read.

The simplicity of science has practical advantages. Ortega y Gasset (1929) wrote: "Contemporary science can put blockheads [*tontos*] to good use".

The mismatch, between a world or environment that is complicated and a science that is simple, is exposed by the obfuscating jargon so often used to describe the environment. Herbert Read (1949) wrote, "Almost any standard work on psychology or sociology will provide an example [of jargon] that is not a parody". He added that the characteristic of the great English stylist such as Shakespeare is to use the wealth of transitive verbs afforded by the language. To avoid plain verbs sometimes seems like a principal skill of the social scientist.

B.F. Skinner, the great behaviourist (now more widely cited even than Freud), complained trenchantly of the pathology of gobbledegook: "I accuse cognitive scientists, as I would psychoanalysts, ... of inventing explanatory systems which are admired for a profundity which is more properly called inaccessibility". Personally, I find Skinnerism, with its "schedules of reinforcement", on "fixed and variable intervals and ratios", subject to "contingencies", even more of a private language than the branches its founder criticises.

There are other pathologies within Psychology, caused by workers trying to grapple with a subject matter of inordinate complexity. For one thing, we have a situation where,

> Practically every theorist and almost every new doctoral candidate tries to build a new and different model of personality, rather than trying to iron out the anomalies and inconsistencies accumulating in existing theories by the processes of normal science.
>
> H.J. Eysenck (1988)

In other words, in practice scientific analysis breaks down when it comes to complex phenomena like personality.

For another thing, we have the pathology of obsessive intellectual imperialism, where a worker tries to deal with complexity by shrugging off everything outside his own fixed approach. Freud himself said, "Psychoanalysis became the whole of my life".

Finally, we have the syndrome of problem avoidance. Problem avoidance is central to Lakatos's picture of science, since "research programmes" offer the scientist guidance about both which research problems to avoid and which to pursue. The former are the "negative heuristic" of the programme. But in Psychology this negative heuristic seems to include anything interesting or fundamental, in fact anything at all that the non-professional would call "psychology". It excludes anything involving the real environment in its rich but baffling complexity.

"SOCIAL SCIENCE"

Expressions like "bitter sweet" are called oxymorons. The term "social science" may be an oxymoron, science being appropriate only for systems simpler than society. I should say, too, only for systems simpler than the human participants in society.

This stipulation would be unnecessary on a sufficiently loose Feyerabendian definition of science. Bridgman was a Nobel physicist best known for emphasising the importance of "defining operations" in science. He once said: "The scientific method is doing one's damnedest, no holds barred" (1959). Anyone can do their damnedest to understand society.

On the other hand the great physicist Marie Curie held that "Science deals with things not people" (Giroud, 1986). Her remark raises a legitimate anxiety whether attempted science of people will deal with them *as if* they were only things. There are many obvious contrasts between the natural and the social worlds. For example, Karl Mannheim, who originated the sociology of knowledge, regularly refers to the natural world, and to the concepts appropriate to its study, as "timeless and static". For Mannheim, then, the far from timeless or static social world would not be amenable to the same sort of study.

In this book I am studying a conflict of ideas between environment and heredity. Many such conflicts, and I shall argue in Chapter 7 that this is no exception, seem to boil down to politics, and to one's whole approach to history.

Long before modern Psychology began, the question was debated: is history a science? The claim that it is certainly did not originate with Marx. For Vico, in his 1725 *Scienza Nuova*, history is the *greatest* science. Renan, too, in his *Life of Jesus* (1863), has it that "History is a science like chemistry, like geology". None the less it is Marxism, with its enormous political influence on the Soviet Union and China, whose claim to be scientific matters most. How many people's lives have been blighted by the view that "genuine scientific knowledge of the laws of the historical process leads with irrefutable iron necessity [to certain political conclusions]" (Hessen). It is because science, now, seems to bestow the certainty that was once reserved for religion, that Marxism has been called a Christian heresy.

The lure of "irrefutable iron necessity" has to be reckoned with in any field. When we turn to Psychology itself, we find that the winning slogan chosen for the American Psychological Association's 1992 centennial was "a century of science and service".

Some say that Psychology is just the study of the brain, and as such another science of matter. Since Descartes, matter has been defined as that which occupies space, while mind is the "I think". The language of mind is different from the language of brain, and it is wrong to confuse them. Thus, Brentano (1874) held that everything mental is "intentional", directed towards something other, unlike matter. The philosophy of the relation of mind and brain goes back to the ancient Greeks, and Aristotle expressed the resilient position known as the dual aspect theory. This views brain and mind as two different aspects of the same underlying thing. This is Spinoza's (1665) position (and mine).

The idea that Psychology is another science of matter goes with a hereditarian leaning, since the understanding of the bodily mechanism of genetic inheritance has enjoyed major recent advances. Much traditional philosophy can be seen to bear on the question of environment.

The branch known as "artificial intelligence" (AI) has given new life to scientific Psychology, but on Brentano's (1874) view the term is another oxymoron. The AI school hold to the "Turing test", that a machine is "intelligent" if it behaves indistinguishably from an intelligent organism. The contemporary philosopher Searle (1992) has given an argument to show that this is not a valid test.

I have already mentioned Gergen's (1973) assault on the idea of *Social* Psychology as science. Let me outline further respects in which he holds his own discipline to differ from science.

He says, "The recipient of [Social Psychology] is . . . provided with dual messages: Messages that dispassionately describe what appears to be, and those which subtly prescribe what is desirable". For example to be categorised an "introvert" is an implicit prod to become more outgoing.

In a similar way, Gergen puts Skinner's "scientific Law of Reinforcement" into a common sense perspective:

> Parents are accustomed to using direct rewards in order to influence the behaviour of their children. Over time, children become aware of the adult's premise that the reward will achieve the desired results and become obstinate.

Gergen makes a third telling point. The goal of science is often said to be prediction, yet, "In the same way that a military strategist lays himself open to defeat when his actions become predictable, an organisational official can be taken advantage of by his inferiors and wives manipulated by errant husbands when their behaviour patterns are reliable".

Schlenker (1974), in his rebuttal, argues that this very "poverty of predictability" is a transhistorical truth of the kind that Gergen denies are to be found in Social Psychology. To me this sounds like saying that the only transhistorical truth in Social Psychology is that it isn't a science. Schlenker has described Gergen's position as being that the innate/instinctual can be scientifically explained, while the learned and therefore cognitively modified cannot.

Social (including Environmental) Psychology is perhaps the hardest branch to regard as science, but the issue arises throughout the discipline, as I discussed in *Psychology on the Couch*. Freud had a controversial idea that women from early girlhood feel inferior to men. He called this "penis envy". Early psychologists developed what might be called "physics envy", believing they could only be respectable if they modelled themselves on physicists.

The founding father, Wundt himself, acknowledged that there is an unscientific *part* of Psychology and wrote his huge *Völkerpsychologie* (Wundt, 1916) to make a start on it. Early Psychology in America on the other hand, at least as represented by Titchener, discarded this side of Wundt's ideas. The later American, Skinner, has been unequivocal that: "The ease with which mentalistic explanations can be invented on the spot is perhaps the best gauge of how little attention we should pay to them . . . It is science or nothing".

To me, he seems to be saying no more than that "it's all too complicated". My own view is much closer to the one that "Our forms of life lie beyond science, but not beyond disciplined inquiry" (Leahey, 1987: 471). This view is much more sympathetic to alternative approaches such as the "hermeneutic" one. Freud said that dreams can be interpreted, and wrote a book to say how. But perhaps we can regard not just dreams but all of behaviour as texts to be scrutinised for meaning (which is the hermeneutic approach).

It has been claimed that Clinical Psychology students mostly regard the scientific part of their training as a boring chore. George Albee, in particular, argues that it was a mistake, to begin with, for clinical psychologists to model themselves on physicians, when what they should be is agents of widespread social change. This is now the major battleground for whether Psychology should be treated as science. In America, between 1960 and 1980, the Psychology labour force grew 475%, and by 1980 applied psychologists made up about 61% of all doctoral psychologists, while traditional experimentalists were only 14%. Also, in 1967 62% of new doctoral psychologists took work in colleges or universities, while by 1981 the figure was down to 32%. Between 1957 and 1976, the percentage of Americans who had consulted a mental health professional rose from 4 to 14%; among the college educated, the change was from 9 to 21% (Leahey, 1987: 477f). The tide of the discipline is flowing towards those who regard its scientific part as a boring chore.

I believe the academic publication process militates against valid consideration of the environmental outlook. Prolific academics, "success stories", prefer to believe their success came from internal causes rather than the circumstances of external environment. I expand this argument in Chapter 7.

In this light, it is disturbing that the road to publication is made easier for those already well known. The supposed protection against this—"blind" reviewing—is a misnomer (Kupfersmid, 1988): the author's identity can usually be guessed anyway. A study by Peters & Ceci (1982) showed that articles copied from leading journals, retyped and resubmitted to the same journals under a different (less prestigious) name/ affiliation, were not recognised and were rejected for publication. This contradicted a claim of Zuckerman (1978) that institutional affiliation does not significantly affect editorial and referee decisions.

"Editorial power" is the likelihood of a peer reviewer having their recommendation on publication accepted by a journal editor. This, too, is greater for the prolific, though not for the highly cited, who tend to be over-critical (Kupfersmid, 1988). Other criticisms of the peer review system are discussed by King (1987). For example, it is discouraging that what little agreement there is, between separate reviewers, is due to agreement on a list of "don'ts", i.e., characteristics leading to rejection of a manuscript, rather than to agreement on desirable positive characteristics. Any author who has had a manuscript rejected will have feelings about the publication system, but there is more to it than that. We live in an age when microchip technology is opening up academic communication, and it will have a great impact on received wisdom.

ANNOTATED CHAPTER BIBLIOGRAPHY

Bacon F (1620). *Novum Organum.*

Barnes B (1985). *About Science.* Oxford: Blackwell.
An interesting book by an author who has done much to put "science studies" on the map in the UK.

Ben-David J (1962). Productivity and academic organisation in nineteenth-century medicine. In B Barber & W Hirsch, *The Sociology of Science.* New York: Free Press.
Just one of the author's many contributions to the sociology of science.

Berman M (1984). *The Re-enchantment of the World.* New York: Bantam Books.
Morris Berman has written few books—it would be good to see more.

Brentano, Franz (1874). *Psychology from an Empirical Standpoint.* 1981 edition, London: Routledge & Kegan Paul.

Bridgman PW (1959). *The Way Things Are.* Oxford University Press.

Cohen IB (1985). *Revolution in Science.* Cambridge, MA: Harvard University Press.
A kind of Darwin to Kuhn's Wallace.

Dubos R (1965) Science and Man's Nature. *Daedalus* **94**, 223–244.
The author is a distinguished life scientist with a gift for self-expression and many interesting ideas, a French emigré to the US quite early in his long career.

Eysenck HJ (1988). Theories of personality. In AW Staats & LP Mos *Annals of Theoretical Psychology 5.* New York: Plenum.

Feyerabend P (1975). *Against Method.* New York: Free Press.
A book that infuriates some people. I remember seeing the great iconoclast, giving an address, while waving his stick, as though he would like to hit Method with it.
Gergen KJ (1973). Social psychology as history. *Journal of Personality & Social Psychology 26,* 309–320.
Not many journal articles are worth the time of the non-specialist. This is one. A humanistic psychologist tackles the scientific orthodoxy. The thesis is so powerful that I wonder it is not better known.
Giroud, Françoise (1986). *Marie Curie: A Life,* trans. by L Davis. London: Holmes & Meier.
Heilbron JL (1986). *Dilemmas of an Upright Man: Max Planck as Spokesman for German Science.* San Francisco: University of California Press.
Heather N (1976). *Radical Perspectives in Psychology.* London: Methuen.
Heather is a clinical psychologist, and speaks for many in that profession who find their academic studies a disappointing preparation for the work they do. But this short book fails to convince me personally that radical perspectives are worth pursuing.
Henle, M (1986). *1879 and All That: Essays in the Theory and History of Psychology.* New York: Columbia University Press.
The fruit of a long career in Psychology which has left a healthy disrespect for Wundt.
Jaki SL (1978). *The Origin of Science and the Science of its Origin.* Edinburgh: Scottish Academic Press.
Shelved in the Religion section of my library. As the title implies an odd, though intriguing, mixture.
King J (1987). A review of bibliometric and other scientific indicators and their role in research evaluation. *Journal of Information Science 13,* 261–276.
A first-rate scholarly review. The computer story has meant much more "bibliometric" data, especially on citations, is now available. So long as proper caution is exercised in the interpretation of it, I believe it can do a tremendous amount to improve research in any field.
Kuhn TS (1962). *The Structure of Scientific Revolutions.* Chicago: University Press.
A book which changed the philosophy of science irrevocably. Mesmerically readable, the author confesses how much it cost him personally to write: it should be one twentieth-century book that is remembered.
Kupfersmid J (1988). Improving what is published: a model in search of an editor. *American Psychologist 43,* 635–642.
The editor hasn't appeared yet. But some at least are aware of the need for change. The foundation editorial by Estes of *Psychological Science,* the flagship journal of the new American Psychological Society, is remarkable in its trenchant frankness about the journal system. He believes journals exist for the CVs of academics and for the profit of publishers, and that virtually nobody reads them. None the less, he's heading up a new one!
Lakatos I & Musgrave AE (1970). *Criticism and the Growth of Scientific Knowledge.* Cambridge: Cambridge University Press.
Although picking a middle way between Popper and Kuhn, Lakatos, who fled Hungary after the 1956 uprising, held fast to the rejection of fake authority which is implicit in viewing science as a human enterprise.
Leahey TH (1987). *A History of Psychology, 2nd edn.* Englewood Cliffs, NJ: Prentice-Hall.

Not as chunky as Boring or Brett, but fully comparable with them. Not everyone realises that history, when it means our *perception* of the past, is constantly changing.
Lindeboom GA (1975). *Letters of Jan Swammerdam to Melchisedec Thevenot.* New York: Swets.
Macaulay TB (1843). *History of England from the Accession of James II.* 1979 edition, H Trevor-Roper, Harmondsworth: Penguin.
Merton RK (1965). *On the Shoulders of Giants: A Shandean postscript.* New York: Free Press.
Mulkay MJ (1979). *Science and the Sociology of Knowledge.* London: George Allen & Unwin.
Needham J (1954–). *Science and Civilization in China.* 7 vols. Cambridge: Cambridge University Press.
Ortega y Gasset J (1929). *The Revolt of the Masses.* 1964 edition, London: WW Norton.
Peters DP & Ceci SJ (1982). Peer-review practices of psychology journals: the fate of published articles, submitted again. *Brain & Behavioral Sciences 5,* 187–255.
This journal uses "open peer review", that is, it prints the comments of referees after the article. I think this is basically an exciting idea, but unfortunately far more reviewers than the usual one or two are consulted, and there is prior anonymous reviewing *as well.* Consequently the articles tend to be "block-busters", which is exactly the syndrome academics should be trying to eradicate.
Popper KR (1963). *Conjectures and Refutations.* London: Routledge & Kegan Paul.
It was a major insight that falsifiability is more important to science than verifiability is, but Popper held back from any assault on the objectivity of science.
Read, Herbert (1949). *The Meaning of Art.* Harmondsworth: Pelican.
Renan, Ernest (1863). *Life of Jesus.* 1935 edition, London: Watts.
Richards S (1983). *Philosophy and Sociology of Science: an Introduction.* Oxford: Blackwell.
A useful textbook, but no substitute for Popper, Kuhn, Feyerabend and Lakatos in the original.
Roberts, JM (1985). *The Triumph of the West.* London: Guild Publishing.
The book of a television series. The author is a very distinguished historian who has been Vice-Chancellor of Southampton University and Warden of Merton College Oxford. His most ambitious project has been a one-volume History of the World.
Schlenker BR (1974). Social psychology and science. *Journal of Personality & Social Psychology 29,* 1–15.
Somebody had to make the orthodox case against Gergen's views and this is it, done well, or at least as well as it can be.
Searle JR (1992). *The Rediscovery of Mind.* Cambridge, MA: MIT Press.
Spencer, Herbert (1876—1896). *Principles of Sociology.* 3 vols. 1969 edition, London: Macmillan.
Spinoza, Baruch (1665). *Ethics.*
Staats AW (1991). Unified positivism and unification psychology: Fad or new field? *American Psychologist 46,* 899–912.
Weber M (1946). *Essays in Sociology.* Oxford: Oxford University Press.
A modern reprint of the work of this great figure in intellectual history.
Westfall RS (1981). *Never at Rest: A Biography of Isaac Newton.* Cambrige: Cambridge University Press.
Williams SM (1988). *Psychology on the Couch.* Brighton: Harvester.

I don't take back a word of it.
Woodworth RS & Schlosberg H (1955). *Experimental Psychology*. New York: Henry Holt.
The bible of mid-century Psychology.
Wundt W (1916). *Völkerpsychologie. Elements of Folk Psychology*. Translated by EL Schaub. London: George Allen & Unwin.
Young RM (1970). *Mind, Brain and Adaptation in the Nineteenth Century: Cerebral Localization and its Biological Context from Gall to Ferrier*. Oxford: Clarendon.
This wonderful book on phrenology showed me in fine detail, as I reached the stage of disillusion with my early research, which was in "hemispheric specialisation", how much history repeats itself. Yet both phrenology and hemispheric specialisation have enjoyed the most amazing popularity—if it were spelled out why, we should understand the neglect of environment much better.
Zuckerman H (1978). Theory choice and problem choice in science. In J Gaston, *Sociology of Science: Problems, Approaches and Methods*. London: Jossey-Bass.
The author is best known for a study of American Nobel laureates which gives much insight into the psychology of science.

CHAPTER 2

The Rise of Environmental Psychology

PART A. METAPSYCHOLOGY OF THE ENVIRONMENT

The strand changes, lapping or pounding,
The waves of the other do this to the self.
Williams (unpublished)

"Metapsychology" is, analogous to "metaphysics", upon or about Psychology rather than actually being Psychology. In new branches of Psychology, such as this one of Environmental Psychology, there is less need to continue a historical development. So a meta-perspective, to recognise where we are, and map out a future, can be particularly useful. Though "environment" has the distinction of *not* making it into Raymond Williams's compilation of *Keywords: the Vocabulary of Culture and Society*, it is unquestionably a genuine keyword of the late twentieth century.

The formation of Environmental Psychology has been unco-ordinated, with many streams flowing into the river. Some writers have adopted the pessimistic approach to its definition, of falling back on "what environmental psychologists do". One textbook definition captures important emphases, "the study of the inter-relationship between behaviour and the built and natural environment". Gilgen's (1982) profile of *American Psychology since World War II* describes Environmental Psychology as one major force within the contemporary discipline.

In some ways, it is surprising that the term "environment" had to wait so long to be linked with "psychology". For the marriage usefully conveys the emergence or intensification of two major concerns.

THE GREENS

The first of these concerns will play the minor part in this book. It is the concern with the worship of economic growth.

For many years the worry of economists has extended to intense *public* concern about inflation and unemployment and, to a lesser extent, weak currency and balance of payments. To all these ills, people propose the panacea of growth in economic output. Yet, repeatedly, the environmental lobby has pointed to examples of industrial expansion degrading the environment. The power stations belch oxides of nitrogen and sulphur into the atmosphere, there to acidify the rain and so destroy life. Nuclear waste is dumped at sea as a slow time bomb. Research has identified a problem of global warming, caused by carbon dioxide released into the atmosphere by industrial processes. Chlorofluorocarbons (CFCs), used in refrigeration and aerosol sprays, are destroying the protective ozone layer of the atmosphere.

The weaker a country's economy, the greater, generally, the pressure for growth, and so the greater the resistance to those who insist that the environment should be clean. Since acid rain, for example, knows no frontiers, there is an inherent international dimension to environmental awareness. Visiting Eastern Europe, the levels of pollution, accepted as a matter of course, have appalled me. I vividly remember driving through Jesenice in Slovenia, when a foul orange industrial smog choked the whole town, drastically reducing visibility. Will a growing consciousness of one Europe help to lessen such blight?

The motivation of entrepreneurs is profit, and it would be unrealistic to expect this to be readily deflected by environmental considerations. So there is a need for comprehensive cost–benefit analysis of economic expansion, which is most easily undertaken by public organisations or by government itself. There is a perception that environmental concern is at least in tension with traditional capitalism. There is a spreading belief that we are stewards of our physical environment, even on a planetary scale. Together, these beliefs have fostered new political parties in several countries. Now generally adopting the name "Green", some of them have achieved startling electoral prominence in a short time. Now, they may be settling down to a consolidation phase. I feel, for my part, that a political philosophy centred on the environment should be careful not to neglect the needs of the poor. It is the positive effects of economic growth that in the past have done most to meet these.

BORN OR MADE?

The second service to Psychology of the term "environment" is its connection with explanations in terms of personal experience. We seek the causes of behaviour in the individual's interaction with his environment. Since Freud we have been aware that it is not merely the current environment that matters, but also the environment during earlier development. Indeed Freud argued that, because, during infancy and childhood, we are unable to cope so well with threats from the environment, they can act as "traumata". Traumatic experiences persist in disturbing our lives, though we may be unconscious of them, and so they may be even more important than the current environment. In this sort of sense, people often oppose "environment" to "heredity" as a determinant of behaviour, the old nature/nurture debate. This second sense of Environmental Psychology is the fundamental one for this book, seeing the ecological movement as a way of enriching our notion of "nurture".

This is far more than a purely academic controversy. A good example of this is Singapore during the long rule of Prime Minister Lee Kuan Yew. Lee is on record as fearing an impending "genetic collapse" of his country. He accepted totally the traditional consensus of "behaviour geneticists" (see Chapter 6) that heredity is four times as influential on intelligence as environment. Research showed many female Singaporean graduates were remaining unmarried. Two-thirds of male graduates were marrying beneath their educational level. He felt they should marry at their own educational level (he saw a role for computerised matchmaking). Graduates were having fewer children than those of lower educational level, and he exhorted them to have more. Otherwise, their country would collapse like Classical Greece, where, according to a genetics lecturer at the National University of Singapore, self-indulgent educated women pursued worldly pleasure at the expense of motherhood (see Gould, 1985). Perhaps he felt women had emancipated themselves too freely since the abolition of polygamy in Singapore in 1960. Some conservatives in his country were saying this had opened a "Pandora's Box" of women who are selfish, career-obsessed, money-mad and tainted by Western values. They said that "so-called 'women's liberation' is nothing but a lesbian conspiracy". In 1984 Lee took action: he gave graduate mothers first preference for primary schooling of their children. He gave next preference to those non-graduates who had agreed to sterilisation after their first or second child. Singapore is no backwater, rather one of the most rapidly expanding economies in the world. In Chapter 7, I shall go on to broaden my discussion of this "eugenic" philosophy.

In Britain, politicians are not so blunt. Yet there exists a service of "genetic counselling", to apply the discoveries of genetics to the prevention of physical disease. This obviously has potential to be misused to discourage the poor from having children. I find the absence of comment on this disturbing. It created a stir once, when newspapers reported that the Indian government was offering free transistor radios to people who would accept sterilisation. Could this be classed as genetic counselling?

In the area of mental disorder, too, the controversy has important implications for social policy. People use the genetic view, often elaborated into biochemistry or neurology, to justify the involvement of law, i.e. "sectioning" and compulsory treatment, in the mental health field. I have something to say that will seem to some extravagantly cynical. I cannot help feeling that the hereditarian standpoint acts as an excuse for practitioners to "leave well alone" or, if you like, "put our feet up", since nothing can be done for the genetically deficient. In my view, historical progress largely consists in a recognition that we are *not* powerless, and that environmental contingencies do have an effect. For example, the old term for schizophrenia was *dementia praecox,* with a prognosis of irreversible deterioration until death. The rise of the new term signified, at the time, greater optimism that correct treatment could at least stabilise the condition.

EMERGENCE OF THE DISCIPLINE

Given these external social pressures for the formation of an Environmental Psychology, what general features characterise the fledgling discipline?

There is a tendency towards a systems approach, with the study of environment and behaviour as an integrated unit. More important, is an inclination to applied research (sometimes called a "problem-orientation"), rather than theoretical or basic work. Public criticism of traditional Psychology frequently focuses on its excessively "pure" nature. Paradoxically, however, the public often reject applied Social Psychology, when confronted with specific examples of it: such rejection is a cross that Environmental Psychology, too, must bear. Another of its characteristics is its multi-disciplinary appeal. Some interested occupational groups include architects (including landscape architects), planners, builders and industrialists. Cognate academic areas include urban sociology, urban anthropology and recreation studies. A further characteristic is a reaction against traditional methodology, on grounds I discussed in *Psychology on the Couch* (Williams, 1988). These workers emphasise especially the artificiality of laboratory Experimental Psychology as well as its concentration upon measures taken through a short time span.

Historically, then, how did Environmental Psychology arise? Perhaps a better question to ask is, why did it take so long? Dubos (1972) confirms the analysis I began in the previous chapter,

> The study of man as an integrated unit and of the ecosystems in which he functions is grossly neglected because it is not in the tradition which has dominated science since the seventeenth century.

Within Psychology, a prominent early sympathiser was Egon Brunswik. He pointed out that, if we really wish to find valid laws, we must pay as much attention to an adequate sampling of the aspects of the environment as to an adequate sampling of individuals. I see the Gestalt movement as

Figure 2.1 High-rise. Reproduced by permission of Mirror Syndication International

a precursor in some ways. Thus, it attacked Wundt's (1912) psychophysics and Ebbinghaus's (1908) associationism, because of their unrealistic "atomism". Also it emphasised the "ground" (or environment) within which one views a figure. So, too, can Lewinian field theory be seen as a precursor, with its formula $B = f(P, E)$ (behaviour is a function of the person and of his environment). It was children that Lewin (1951) studied, in part because this simplified total monitoring of environment (as well as of behaviour). A similar methodological concern can perhaps be seen in the study of single cases so important in "Behavioural Analysis", sometimes called the "Experimental Analysis of Behaviour" (another modern advocate of "$N = 1$" is Herbst, 1970). Lewin's idea of "atmosphere" (e.g., authoritarian vs democratic) is also environmental. It also should be said that the successes of traditional Psychology in gaining public acceptance have given greater confidence to enable a tackling of the real environment. In the world outside universities, the requirements of post-war reconstruction may have given the major impetus. Later this shaded into the problem of urban blight.

In the 1950s, the academic field took shape with the birth of three principal areas of study. One area was behavioural investigation of environmental design, focusing initially on psychiatric wards. Confirmation that design critically affects nurses' access to patients emboldened more ambitious projects. In many countries the desire to economise on expensive development land led architects and planners to the high-rise idea (Fig. 2.1), the social inadequacy of which environmental psychologists were later able to document. Like medicine, architecture has a language of its own, from architraves to Brutalism to cornices, and it was an achievement of environmental psychologists to break the code (Fig. 2.2). I shall consider architectural design much more in Chapter 5. The controversy that followed the description by Prince Charles of a new building as "a monstrous carbuncle on the face of London", gave a good insight into the feelings it stirs.

A second area, first dubbed "proxemics", studied people's needs for personal space and territory. Anthropological studies by Hall (1959) led forward much-reviewed work on cross-cultural, situational and individual-difference determinants of personal space. Modern investigations have also looked at interactions with the physical environment (e.g., effects of low ceilings, room corners, no light, being outdoors) and at effects of friendship with the personal space "invader". One interesting research programme concerns the utility of territorial boundaries, such as the Peace Line in Belfast, in diminishing aggression.

Proxemics is also related to a corpus of work on the effects of crowding. Appalled by the contrast between the world population back in evolutionary history and today's rapidly growing billions, many today believe we have a population problem. There have been studies of animal samples

Figure 2.2 The language of architecture: mosques in Istanbul

breeding through successive generations within a fixed territory, where other environmental conditions such as availability of food and water are fine. These reveal the development of what is called the "behaviour sink", which has foreboding implications. The historical drift towards cities exacerbates the anxieties (one-third of the UK population lives in seven great conurbations).

There are some who would deny the relevance of animal studies. One ground would be that the environment of an animal species remains constant, in the sense that the activity of that species upon it does not alter it. Man however makes his own environment. He can adjust his environment rather than having passively to adapt to it. Adjustment rather than adaptation has become more possible with historical progress, as even rather trivial examples like central heating show.

The mention of cities leads into the third early area of Environmental Psychology: "urban imagery". This term refers to work by, for example, Lynch (1960), asking subjects to draw maps of their home town or perhaps of some smaller neighbourhood. These "cognitive maps" reveal predictable individual differences, probably reflecting different amounts of experience with the relevant environment. To give a flavour of the sort of research: they show that middle-class people think they are further away from slums than they really are. Later work by Gould & White (1974) coloured in cognitive maps of whole countries in terms of the perceived residential desirability of particular regions. Another direction taken by later work was to explore the general effects of living in cities. People highlighted the urbanite's highly planned, deliberate style; his greater tolerance of many forms of deviancy; the association between commuting and high blood pressure. Some of this work depended upon global urban/rural comparisons. Other aspects studied single variables: physical stressors such as noise and air pollution (another such stressor is weather).

These three areas, then, established Environmental Psychology as a force in the 1950s. Also, competing descriptions such as "Architectural Psychology" and "Man–Environment Relations" seemed to fall away. The 1960s saw the founding of the first journals (three of them), the publication of "classic" programmatic articles, conferences in the field, and the founding of specialised research institutes. The 1970s witnessed the establishment of the American Psychological Association's Division 34 for Population and Environmental Psychology and the mushrooming of college programmes.

THE SHAPE OF ENVIRONMENTAL PSYCHOLOGY

As to the general shape of modern Environmental Psychology, a few comments from a meta-perspective are appropriate.

With any new branch, there is a temptation to cut loose from the past, but Environmental Psychology has striven to maintain organic links with the tradition in two topics particularly. Our immediate interface with the environment is perception, and here there is a task of selecting the most relevant parts (such as the work of Gibson, 1979) of the corpus. The topic of adaptation to new environments (including its potency against cumulative effects) can be illuminated by traditional work (Kohler & Wallach, 1944) on "figural after-effects".

Another point concerns the relationship of political questions to Environmental Psychology. Many of the issues are so important and charged that politics is inevitably involved. At the same time politics is a behavioural arena that particularly interests us as psychologists. For example, one issue is noise pollution and the possible need for legislation to combat it, when one study showed city-centre noise increasing by more than 10 decibels over a decade. Another issue is environmental lead, and its implications for IQ, where pressure groups throughout the West waged a campaign to secure controls on lead in petrol. Another example of conservation entering politics is the public anxiety aroused by proposed new nuclear power stations, such as Sizewell B in Suffolk, England. The planning inquiry received much publicity, but only showed the financial difficulties of opposition to the giant corporation. Politics also shades into the study of the psychological effects of disasters such as earthquakes, volcanic eruptions, hurricanes and tornadoes, and major industrial accidents like Bhopal, Seveso and Chernobyl.

The range of topics now catalogued shows that Environmental Psychology must espouse an eclectic methodology, including theoretical work, and people within the field explicitly recognise this fact. Naturally, it remains the challenge of the value coherence to demonstrate the interconnections of the various topics. No doubt, over time the field will develop greater uniformity. Perhaps an extreme of the present range is the interest shown in the history of Eastern (oriental) attitudes to the environment. One reason for exploring every avenue in this way could be a centrality of attitudes and evaluation to Environmental Psychology (evaluation is clearly linked to selection for perception).

PART B. THE PHYSICAL ENVIRONMENT

Environmental Psychology has begun modestly with a concentration upon the physical environment, but the promise (or threat) is there to broaden the frontiers to take in the social environment. The purpose of this chapter is to sketch in some subject matter of what can, alas, already be called traditional Environmental Psychology, i.e., the Psychology of the physical environment.

It is striking to note the evolution towards environmentalism within the thought of one very prominent psychologist, Ulric Neisser. He first came to that prominence through the publication in 1967 of his book *Cognitive Psychology*. This synthesised a great range of mainly experimental data on internal psychological processes. It acted to stimulate the development of a large branch of the field of Psychology, with the same name as that of his book. By 1976 Neisser himself, at least, was coming to recognise that generalisations about cognition that do not consider the environmental context are risky. For in that year, he published another book entitled this time *Cognition and Reality*. This shift parallels major philosophical debates, such as whether the foundations of mathematics stem from the nature of mind itself and not from external reality.

SOME THEORIES

Mention of philosophy takes us on to the general role of theories of the effect of the physical environment. Textbooks present analyses in terms of "behavioural constraint", "adaptation level" and "environmental load" well. I should like to single out for particular mention the ecological Psychology of Roger Barker with its theory of the "overmanned" or undermanned environment. Barker (1968) believes the best predictor of human behaviour is where someone is: "In the post office we behave post office, no matter what else we may have on our agenda." Another analysis that I want to relate to recent work depends on the concept of "environmental stress". Jeffrey Gray of the Institute of Psychiatry in London has done work on one means whereby we attempt to deal with stress: the drug Valium. It is now widely known that this drug, like other benzodiazepines, is dangerously addictive. Gray (1985) believes it also actually weakens coping ability, our natural resilience in the face of stress. The message seems that we have what might be called "psychological crutches", and it is best to do without them if possible.

Another theoretical analysis worth mentioning in passing is in terms of "arousal". The virtue of this formulation is that it can predict the way the environment can cause improvements in performance (by remedying under-arousal) as well as decrements (from disorganisation due to over-arousal). One of Psychology's few laws, the Yerkes–Dodson Law (Fig. 2.3), states that there is an optimum level of arousal that one can lie below as well as above. However there is a bugbear with arousal: there seem to be a million and one different possible ways of measuring it. One can use both a variety of physiological indices (e.g., pulse rate, galvanic skin response, palmar sweat, etc.) and self-report indices, and the different indices do not always go together.

ENVIRONMENT PERCEPTION

The immediate interface between the person and his or her environment is the process of perception. The Psychology of perception has filled many textbooks, but what parts of it are especially relevant to the environment?

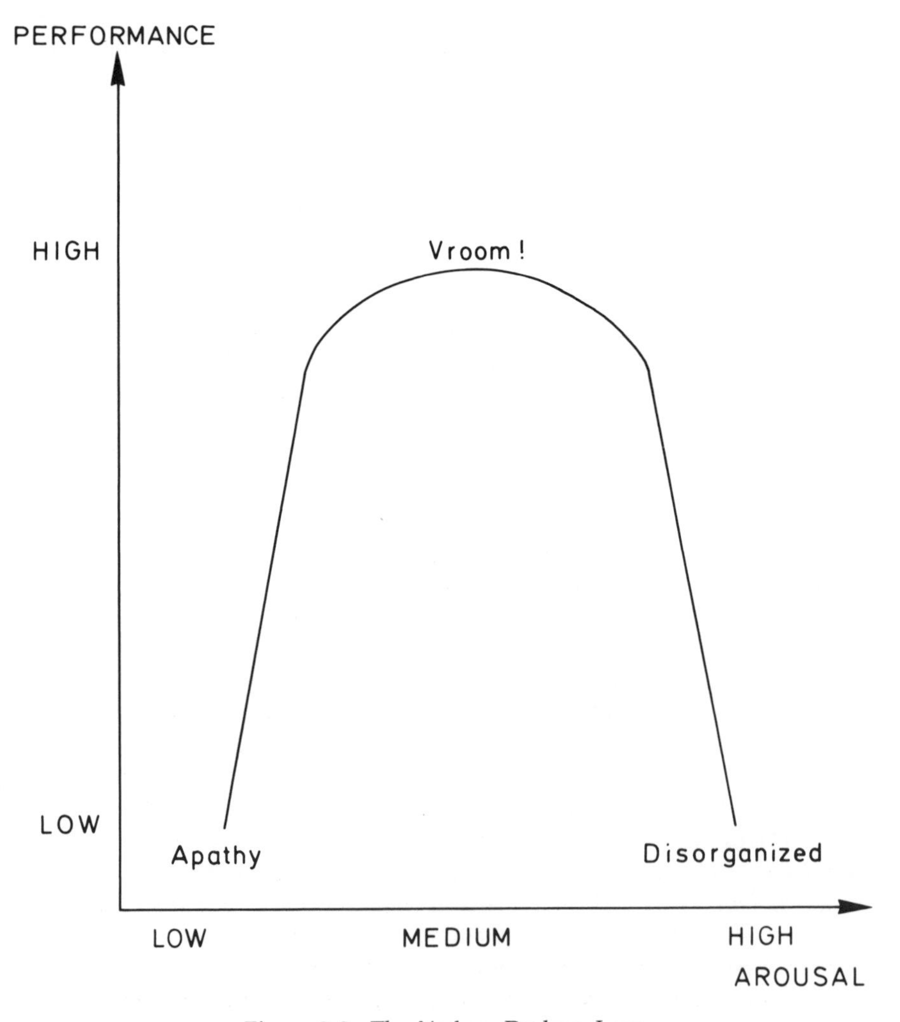

Figure 2.3 The Yerkes–Dodson Law

There is a philosophical problem with perception. Perception can be illusory, e.g. we perceive a stick jutting out of water as bent though we know it is actually straight. So it is common to say, that what we perceive directly are "sense-data", patches of colour and so on, rather than physical objects. It is the sense-data that lead us to the idea of physical objects. But this leads on to a general skepticism about the very existence of a real physical world or environment "out there". Since the physical world is not experienced directly, how can we know about it for sure? One response, to a position that so offends common sense, is known as "naïve realism". This says, "let's start from the reality of the physical world; since the idea of sense-data throws this into question, we'll dispense with them". But the idea of sense-data has seemed indispensable to so many philosophers, that this theory is unlikely to stand up. One other tack, known as "phenomenalism", is to regard statements about the physical environment as completely reducible to statements about all the possible sense-data that could be received from it. But again this fails to deal with an intuitive feeling, that the environment is separate from, and somehow produces, sense-data. What may be closest to the truth is some sort of causal theory of perception: in perceiving, we have sense-data that are caused by what we call physical objects. The objects may therefore be nothing at all like the sense-data. A cute way of putting this theory, is that the physical environment is "in drag" when it is perceived.

I move on to empirical research. The classic work of von Senden (1960) bulks large. He studied patients who had been blind for many years, until surgical relief of their eye cataracts restored vision, long after the normal age for development of the visual powers. In one striking anecdote, he relates how health staff thought one such patient had become unhinged by the sudden opening to him of a new sensory world. The patient had thrown himself from a first-floor window in the hospital. So far from being unhinged, von Senden showed that this man merely needed more time to learn how to use visual information. He had thought that the window in question let down immediately on to the ground—he still had residual deficiencies in depth perception. Von Senden's work thus showed the virtue of taking in the whole person, characteristic of environmental work. It also showed the importance of learning and of interaction with the environment in the development of human capacities.

In this latter respect it contrasts with much modern work with neonates, for which reports by Fantz were seminal. He monitored the eye movements of newborn babes exposed to their mothers' faces. He found an "inbuilt" predisposition of the neonate to look at the mother's eyes. But suppose that these babies had had some experience with the external world before his experiments, and this was sufficient to cause the eye-movement patterns. The fact that he observed the movements even in response to a schematic "child's drawing" of a face with eyes, rendered this less likely. So we see the nature–nurture debate arising even with regard to the early processing stage known as environmental perception.

However, the way we perceive our environment has many practical as well as theoretical ramifications. For example, one important and prevalent phenomenon is habituation – the way we cease to perceive often-repeated features of our environment. For example anyone who has suffered and groaned on the way to and from work will attest to the frequency of repetition in road traffic. The phenomenon of habituation to road traffic even when driving in it may be an important clue to the major social problem of road accidents. Somehow, it should be noted, we all develop a "traffic sense" that enables us to survive in traffic without radically overhauling this habituative feature of our perceptual system. The problem of habituation is a major one for any task requiring prolonged vigilance. The working conditions of radar operators, so important in responding to German air raids during World War II, attracted the attention of applied psychologists such as the Mackworths, who studied vigilant performance under controlled experimental conditions (see Mackworth, 1970).

Some work of Gregory (1966) on a classic visual illusion known as the Muller–Lyer illusion (see Figure 2.4) brings out the need for a holistic perspective (i.e., one involving the whole environment) on perception. He put forward a compelling explanation of this illusion in terms of the straight-line built environments that the subjects of illusion experiments typically encounter all the time. If you look at a street corner where the roof is flat, the roof and the pavement create a configuration similar to the left-hand shape. If you look at the corner of the room you are in, the ceiling and floor create a configuration similar to the right-hand shape. In the first case the corner is nearer to you than the other lines, so you know its length is

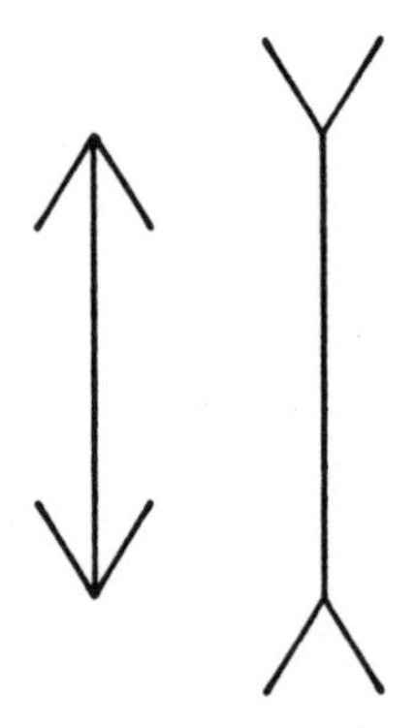

Figure 2.4 The Muller–Lyer illusion

exaggerated relative to them. In the second case the corner is further from you than the other lines, so you know its length is underestimated relative to them. The visual system learns to correct for this. The striking part of Gregory's analysis was that he appreciated that it predicted peoples from other cultures, that do not emphasise stone, concrete and steel building, should not show this perceptual illusion. Sure enough, a study of Zulus in Southern Africa, living in a predominantly natural environment where even the dwellings were not straight-line, revealed they did not experience the illusion.

Context is another word for environment, though a different sense from the ones I have been considering. There is an interesting example, emphasising the importance of context in perception, coming from the study of the perception of speech. It is possible to picture the physical sound corresponding to a piece of speech as a "speech spectrogram" showing changes across time in loudness at each pitch. Though the spectrograms for vowels are fairly invariant in different speech contexts, the opposite is true for consonants. These correspond to rapid changes of pitch in the spectrogram known as "formant transitions". The transition for a particular consonant such as /d/ is completely different depending on the vowel context, i.e., whether the syllable is /da/, /do/, /dee/, etc. Though this is an extension of the word "environment", it assists the general thesis of the insufficiency of atomistic analysis of perception.

THE EARTH'S ATMOSPHERE

One further aspect of the physical environment that standard American Psychology textbooks currently neglect a little is the earth's atmosphere around us. It was right back at the beginning of the twentieth century that Huntingdon hypothesised that the Industrial Revolution itself began where it did because of the temperate climate in north-west Europe. He allied this to grand speculations about high intelligence evolving early there to overcome the cold! He carried this line of thought through to a crypto-racist stereotype about the African "who likes to just sit in the sun". To further detract from his credibility, he speculated avidly about the effects on human activity of the "sun-spot cycle" of about 11 years. He linked it with everything from the business cycle to the circulation of library books.

It does seem likely, though, that the very different climates in the world can explain a good deal about human history. Certainly Europeans living in the tropics are prone to a common kind of intractable lassitude known as "tropical neurasthenia". At the same time, temporary heat waves in industrialised countries are associated to such an extent with outbreaks of rioting that people have dubbed this the "long hot summer" effect. We saw this in Britain most notably in the Toxteth riots in 1982. At the opposite extreme, the all-consuming influence of great cold is most graphically portrayed in Solzhenitsyn's masterpiece, *A Day in the Life of Ivan Denisovitch*. Yet another climatic variable, the wind, has been associated with major psychological consequences. As Shakespeare says in *Hamlet*—"I am but mad north-north-west; when the wind is southerly I can tell a hawk from a handsaw" ("handsaw" was an Elizabethan word for a heron). And only mad dogs and Englishmen go out in the mid-day sun.

The topic of "atmosphere" includes not only climate but also man-made pollution. People have known about pollution from burning coal for a long time. Decrees about brown coal go back to mediaeval times, and Robert Angus Smith coined the term "acid rain" in 1858. It can be shown by studying diatoms (tiny plants) in lake beds that the problem of acid rain started with the Industrial Revolution and accelerated during the rapidly industrialising 1930s. Britain's power stations exude more sulphur dioxide than most European nations but we export our misery: 70% of it leaves the country. In grateful West Germany, half the forests are dying. Even back home people have observed such freaks as rain sourer than vinegar and black snow. Currently, there is also much debate over the effects of CFCs on the ozone layer, and over the "greenhouse effect" caused by an excess of carbon dioxide. There is less disagreement that the mineral lead in petrol has detrimental effects, particularly on the IQ of children. Though now that major national initiatives have been taken to encourage the use of lead-free petrol, further scientific evidence has questioned whether these were really necessary. Certainly since the London winter death rate rocketed in the 1952 smog there has been little resistance in Britain to the idea that atmospheric pollution can be a killer.

There are interesting questions about whether such physical stressors can interact with others—such as intolerable noise—to cause great psychological detriment.

TOWN AND COUNTRY

The central psychological issue about the physical environment however is the contrast between town (*urbs*) and country (*rus*), and this is the issue that I shall explore for the remainder of this chapter. It is a topic of much debate, in everyday life, whether living in cities or in less densely populated areas is preferable. Once, the personal background of those discussing the question would have greatly coloured the debate, for people often spent their whole lives in one type of environment (many still do). There were, and are, fashions pulling people towards one type of environment or other, including, for example, the "back to nature" movement, stimulated by the writings of Thoreau (1854), above all *Walden*. But we live in an age of much greater occupational movement than ever before. The motor car, television and holidays have widened people's experience. Less partial people, with experience of a range of population densities, are now more numerous. The "country lad" (or "lass") is vanishing. Perhaps this has helped the phenomenal growth of detached study of the town/country contrast in modern times. It should be recognised, though, that urbanisation began much earlier in the West than in most of the globe.

Cognitive mapping

One feature of living in cities is the much greater demands it places on one's capacity for "cognitive mapping". This means forming an internal representation of one's environment, to orient oneself and find a way around within it. It is tempting to see cognitive maps as visual imagery. However, the psychologist Kuipers has attacked what he calls the "map in the head" metaphor as insufficient to deal with the realities of the situation. Verbal processes certainly seem to have a lot to do with cognitive maps. Researchers have studied them largely by the simple procedure of asking subjects to draw maps of areas such as that around their home.

They find that any named neighbourhood (such as Mayfair in London) will be pulled towards one's own home. In a study in Cambridge, Terence Lee (1954) asked a sample mainly of housewives to draw maps of their home neighbourhood. He found these neighbourhood maps covered on average an area of about one hundred acres. Their area was determined geographically: population density, which one might expect to be a determinant if a "neighbourhood" is a fixed number of people, had no effect. The concept of a neighbourhood has not been accepted uncritically by all psychologists. They argue that, asked to delineate a neighbourhood, people will do it, but it is none the less, not a real, functioning concept in everyday life. In other words, as in many areas of Psychology, people respond to what they perceive to be the expectations of the psychologist, no matter what is really in their minds.

Cognitive maps can also be studied behaviourally, by seeing how well people can find their way around complex environments. As one might expect, the complexity as well as the lack of distinguishing features of an underground metro such as the London Tube makes it very difficult to form a cognitive map of it.

"Cognitive map" was a term used by the animal psychologist Tolman to reflect his view that organisms learn connections between stimuli ("S–S learning") and not connections between stimuli and responses ("S–R learning"). The S–S vs S–R learning debate has been a prolonged one within the behaviourist school of Psychology. Bruner (1983: 24) has an interesting description of it.

> The first "battle" within psychology of which I was aware was between Thorndike (who interpreted his cats' behaviour in puzzle boxes as blind trial-and-error shaped by rewards and punishments from outside) and Adams (whose proposal was that adaptive behaviour in puzzle boxes grew out of the hypotheses generated by the organism in response to the available cues in the environment). If the environment contained relevant and combinable cues, animals would respond with "insightful" hypotheses. But you could also design blind environments that would make their inhabitants look stupid.

Bruner goes on to discuss the parallel debate, whether learning is passive and incremental, or whether it is stepwise and discontinuous (hypothesis-driven). His own view, is that the continuity view makes man too tamely a creature of his environment, the bad side of behaviourism.

One recent focus of attention, within study of cognitive maps, has been the way in which multi-attribute routes, e.g., roads with many side-turnings, seem longer subjectively than objectively equivalent controls. It also seems likely that cognitive mapping is, in what is becoming a popular phrase, a "right-hemisphere" skill. There is much clinical and experimental evidence suggesting that it is predominantly the right half of the brain that mediates this sort of performance. There has been much criticism of Western education, in somewhat misty terms, that it teaches only the left hemisphere, and neglects important life skills found on the other side of the brain.

Crowding

One most salient distinguishing feature of the city is the extent of crowding in it. Psychologists, so often starting from animal models, have worried about the effects of crowding, from the time it became clear that conditions of overcrowding cause the annual mass suicide of the Norwegian lemmings. Then experimental work of Calhoun (1962) showed that mice colonies confined in a limited space develop major pathologies particularly where the crowding is most acute. This area he graphically described as a "behaviour sink". The behavioural pathologies involved are well known, but one noteworthy finding was that infant mortality was nearly twice as high (which meant virtually complete) in the "behaviour sink". However, other studies subsequently showed that Calhoun's pathological behaviours could be demonstrated in non-crowded conditions.

Yet many worry about the fact that the human world population has expanded from one million in the Old Stone Age to more than five billion today (Livi–Bacci, 1992). Much of the increase has occurred since the

Industrial Revolution, which began with a world population of about 800 million. Life expectancy at birth then exceeded 40 years in no country, while today it is sometimes approaching 80 years. The rest of the population increase was due of course to the development of agriculture, beginning in the Neolithic period. In historical times the greatest increase has been in the Americas and Africa. Periods of great increase, of decrease, and of transition to stagnation can be found in histories such as those of French Canada, Ireland and Japan.

Research studies of human crowding have been many of course. A major technical development has been that "persons per room" has replaced area density in ecological studies of human crowding. The father of ecological studies was the Victorian Charles Booth. He in fact was quite surprised that crowding had such minuscule detectable effects on the health or well-being of his respondents; more recent work by Gove, Hughes and Gall (1979) contradicts Booth (Cochrane, 1983: 71). The existentialist philosopher Sartre wrote "L'enfer c'est les autres" (hell is other people). Is it merely a coincidence that he was a Parisian? Can living in a great city, constantly jostled by other bodies, give one a jaundiced view of the pleasure of their company?

Territoriality and personal space

The problem of crowding in cities is allied to the question of the individual's ability to establish his own territory. Each of us needs a space of our own within which we have freedom of action and privacy. This has been the subject of much empirical research that can be found in textbooks.

Research began with work on animals, such as Niko Tinbergen's studies, which he described in *The Herring Gull's World*. With people, psychologists typically define distinguishable primary territories such as one's home, secondary territories such as one's car or a desk in a library, and public territories such as a towel on a beach.

An old study by Bossard (1931) captures the immense social significance of primary territories. He found that one-third of all marriage licenses issued in a city were given to people living within five blocks of each other.

Secondary territories cause much trouble among British football crowds. The home supporters regard certain areas on the terraces as their own and will defend them against visiting supporters, with regular Saturday afternoon violence. In the British House of Commons there have been some undignified disputes over who takes what seats.

The distinctive character of physical territories is shown by the observation that some areas with a high crime rate retain it through a complete change of population (i.e., through a change of generation).

Often neglected in the empirical literature is the importance of cultural factors. For example some Cambridge colleges are wealthy enough to afford an ordinary room door behind the exterior oak door of student rooms. It is an accepted custom, that when one "sports the oak", and shuts the exterior door, one is deep in path-breaking study, and does not wish to be disturbed.

Another example of a cultural factor, lies in the Jewish Sabbath. The mystic number seven (days of a week) originates in Judaism, though the holy day is Saturday not Sunday. The "Torah" or Law lays down many prohibitions for this day, though, fortunately for the Jews, in the light of the Arab–Israeli wars, fighting in self-defence is now permitted. Even movement is sometimes restricted. The Torah defines four different types of territory, one significance of which is that it is prohibited to transport articles from one type of territory to another on the Sabbath. This prohibition can only be over-ridden by a special rabbinical dispensation called an *eruv*. This law is observed in some American and British Jewish communities as well as in Israel.

Psychologists speak of an invisible territory that we carry around with us, called "personal space". Into this area near our bodies (people often use the metaphor of an invisible "bubble"; Fig. 2.5) others are not welcome to intrude. Katz (1937) introduced the term. Thus there is a characteristic pattern of seating behaviour in public places, such that

Figure 2.5 Our personal "bubble"

someone on his or her own will avoid sitting next to someone else if they can help it.

Naturally, gender affects this sort of behaviour a great deal. For example, close-range interaction is a "tie-sign", signalling that a couple is together. Young children need less personal space. People with "external locus of control" need more. The mentally disturbed tend to need more.

Invading personal space can lead to a "fight or flight" reaction. People walking towards some destination can be deflected by the presence of someone else. The city, again, affords many examples of situations where problems with regard to this basic need arise. The best example is the London Tube in the rush hour. A deliberate depersonalisation of others compensates for the inevitable infringement of personal space. One carefully avoids anything that might be construed as a social signal, any movement including facial movement or shifts of gaze. One often employs "barrier signals", such as crossing the arms in front of the body. We see similar phenomena in the crowded lifts of tall city office buildings.

The topic of personal space is a major one within Environmental Psychology, going back to a book by Hall (1959). He convinced funding agencies that it was an issue of international significance. For there is no doubt that closer approach is acceptable and indeed desired in cultures other than our own. I mean Arab cultures (and probably Mediterranean countries generally), also probably in Japan, with its very high population density (in the Tokyo–Osaka conurbation especially). Hall painted a picture of the diplomatic reception where the Arab ambassador is pursuing the American one around the room. This is not because the American wants to avoid him, but because he is uncomfortable at the close approach of the Arab. As a Briton, I am used to our sacrosanct queues. Visiting Southern Europe, I have been taken aback myself at what amounts to a wild free-for-all to mount the bus when it arrives at a stop: like nothing so much as a rugby scrum.

That is what I want to say about *rus* and *urbs*. Really, the correlated occupational differences, between the farmer and the salesman, say, would be more important, but that would take me far from the physical environment *per se*. In the middle of the great city we find the park, the *rus in urbe*: for how many people has this been the place to go and grieve?

ANNOTATED CHAPTER BIBLIOGRAPHY

Advances in Environment Psychology [Various editors]. Hillsdale, NJ: Erlbaum.
A continuing series of many volumes.

Ayer AJ (1969). *Foundations of Empirical Knowledge.* London: Macmillan.
Greatly superior, according to Casimir Lewy, to his better-known *The Problem of Knowledge.*

Barker R (1968). *Ecological Psychology: Concepts and Methods for studying the Environment of Human Behavior.* Stanford, CA: Stanford University Press.
A book which has had less impact than it might, though Wickens and Gump have been valuable disciples. The Mid-West Research Station set about observing people in their main habitat: supermarkets. "More a method than a theory" (personal communication from Terence Lee).
Bossard JH (1931). Residential propinquity as a factor in marriage selection. *American Journal of Sociology 38,* 219–224.
A study that has lasted in the textbooks a long time, though a behaviouristically inclined student of mine singled it out as an object of aversion.
Bruner J (1983). *In Search of Mind.* New York: Harper & Row.
There is an interview with the author, a very distinguished American psychologist, in Jonathan Miller's videotapes *States of Mind.*
Calhoun JB (1962). Population density and social pathology. *Scientific American 206,* 139–148.
The best research ideas are always simple? Calhoun allowed mice to breed without restriction, with plentiful food and water, in a cage universe where particular areas turned into what he called "behaviour sinks".
Cochrane R. (1983). *The Social Creation of Mental Illness.* London: Longman.
More of this in Chapter 4.
Dubos R (1972). *A God Within.* London: Scribners.
Ebbinghaus H (1908). *Abriss der Psychologie.* Leipzig: Veit.
One of the founding fathers of modern scientific Psychology. Interested in memory, Ebbinghaus realised that forgetting is greatly influenced by the meaning of what is experienced, so that a way to study the effects of time *per se* is to use nonsense language. In the days before college students were willing to advance science, by offering themselves as research "subjects", he learned these syllables, and observed his own forgetting over a long period of time. Another great early psychologist, the Englishman Sir Frederick Bartlett, built his reputation on the same simple idea, though returning to meaningful stories as the "stimuli" learned.
Fisher JD, Bell PA & Baum AS (1984). *Environmental Psychology, 2nd edn.* Eastbourne: Saunders.
The best of the textbooks I keep mentioning.
Gibson JJ (1979). *An Ecological Approach to Visual Perception.* Boston: Houghton Mifflin.
Difficult but influential. A central idea is that what we perceive in the environment are "affordances" such as support or shelter. There is also a good deal on the way in which gradients of "graininess" underlie our visual perception of distance.
Gilgen AR (1982). *American Psychology since World War II: A Profile of the Discipline.* Westport, CT: Greenwood Press.
And really, a profile of the heart of world-wide Psychology.
Gould P & White R (1974). *Mental Maps.* Harmondsworth: Pelican.
Pelican = "very readable", in my experience. This is no exception.
Gove WR, Hughes M & Galle OR (1979). Overcrowding in the home: an empirical investigation of its possible pathological consequences. *American Sociological Review 44,* 59–80.
Gray JA (1985). The neuropsychology of anxiety. *Issues in Mental Health Nursing 7,* 201–228.
Gregory R (1966). *Eye and Brain: the Psychology of Seeing.* London: Weidenfeld & Nicolson.

The classic British text on visual perception.
Hall ET (1959). *The Silent Language.* New York: Doubleday.
Written before "non-verbal communication" filled library shelves.
Herbst PG (1970). *Behavioural Worlds: the Study of Single Cases.* London: Tavistock.
A Norwegian work.
Katz P (1937). *Animals and Men.* New York: Longman, Green.
Kohler W & Wallach H (1944). Figural after-effects: an investigation of visual processes. *Proceedings of the American Philosophical Society 88,* 269–357.
You couldn't want to know more on the topic.
Kuipers B (1982). The "Map in the head" metaphor. *Environment & Behavior 14,* 202–220.
A very stimulating paper.
Lee T (1976). *Psychology and Environment.* London: Methuen.
A very accessible work by the doyen of British environmental psychologists.
Lee TR (1954). *"Neighbourhood" as a socio-spatial schema.* Unpublished doctoral dissertation of the University of Cambridge.
Levy-Leboyer C (1982). *Psychology and Environment.* London: Sage
The view from France. The author has been President of the International Association of Applied Psychology.
Lewin K (1951). *Field Theory in Social Science: Selected Theoretical Papers.* New York: Harper & Brothers.
Livi-Bacci M (1992). *A Concise History of World Population.* Oxford: Blackwell.
As promised, concise. Translated from Italian.
Lynch K (1960). *The Image of the City.* Cambridge, MA: MIT Press.
One of the seminal texts of Environmental Psychology.
Mackworth J (1970). *Vigilance and Attention: A Signal Detection Approach.* Harmondsworth: Penguin.
Neisser U (1967). *Cognitive Psychology.* New York: Appleton–Century–Crofts.
At the time, an astonishing feat of integration of empirical research, with a theme of the individual as a far more *active* contributor to cognition, even at the perceptual level, than had been recognised.
Neisser U (1976). *Cognition and Reality: Principles and Implications of Cognitive Psychology.* San Francisco, CA: WH Freeman.
Solzhenitsyn A (1991). *A Day in the Life of Ivan Denisovich.* London: Harvill.
Roughly as cheerful as his *Cancer Ward.*
Thoreau HD (1854). *Walden.*
Tinbergen N (1990). *The Herring Gull's World: a study of the Social Behaviour of Birds.* London: Collins.
Tinbergen is one of the few behavioural scientists to have won a Nobel prize.
Von Senden M (1960). *Space and Sight: The Perception of Space and Shape in the Congenitally Blind before and after Operations.* (P Heath trans.) London: Macmillan (orig. work 1932).
Williams R (1983). *Keywords: the Vocabulary of Culture and Society.* London: Fontana.
Williams SM (1988). *Psychology on the Couch.* Brighton: Harvester.
Wundt W (1912). *An Introduction to Psychology, 2nd edn.* Translated by R Pintner. New York: Macmillan.
Wundt is generally regarded as the founder of modern Psychology, partly on the strength of having persuaded his university, Leipzig, to give him a room for experiments. In a life wholly dedicated to the new "science", he was far less narrow in approach, himself, than most of his followers. "Wundt studies" are now a thriving academic industry.

CHAPTER 3

The Detrimental Environment: the Question of Evaluation

Das gestirnte Himmel über mir, und das moralische Gesetz in mir.
The starry heavens above me, and the moral law within me.
I. Kant, *Critique of Practical Reason*

INTRODUCTORY

Having said that environmental evaluation is a central topic within Environmental Psychology, it is time to give it some closer attention. I am saying that the environment affects mental health, but what is it that can make an environment detrimental? Like the effects of the physical environment in general, textbooks such as that by Fisher *et al.* (1984) have treated the topic well. Everyone has an idea of what "poor housing" means, for example (Fig. 3.1; 3.2). I want to explore a few specific areas of environmental evaluation in some depth, so what follows is a very personal treatment.

Perhaps the most striking phenomenon with regard to environmental evaluation is the existence of a strong local preference (Gould & White, 1974). We love home, yet it be so basic that others regard it as a sinkhole. This individual partiality is also seen on a larger scale: Gould and White call London "the metropolitan sinkhole" because of the low evaluation non-Londoners give it as a place to live. On this scale there is much evidence in their book for a phenomenon that I shall call "local identification". They show that the locality for which people express a preference is some kind of compact, natural unit.

Local identification, as a form of evaluation, lies at an opposite extreme from purely aesthetic sensations, which are typically highly transient (Goethe once said "Even the loveliest sunset becomes boring after ten minutes"). When an evaluation remains stable over years we talk of an attitude, and attitudes are one focus of this chapter.

Interesting insight into environmental evaluation may be gained from sources other than scientific research, for example, the study of mythology. Thus the word for paradise of the Australian aborigines is "the gum-tree

Figure 3.1 The detrimental environment: The Bogside of the City of Derry, Northern Ireland

Figure 3.2 The detrimental environment: funeral of two policemen in Coleraine, Co. Londonderry, 1987

country": this is fertile, running with water and abundant in game. Similarly, other cultures have the images of the forest cabin, or the valley, or the island, in their metaphors of celestial or earthly paradise.

Historically, there has been a strong ambivalence about the different basic types of environment. People saw the wilderness, on the one hand, as a place of chaos plagued by demons. On the other, it was a source of purity and purification—an idea going back to the sojourn in the wilderness of Jesus of Nazareth. Chapter 2 began discussion of the *countryside*: viewed sometimes as idyllic, but also sometimes as a place of melancholy and grinding peasant life. Similarly, people have seen the city not only as a place of order, freedom and gaiety, but also of worldliness, corruption and oppression.

Some writers, now commonly grouped as "Romantics" (e.g., Rousseau, Blake, Wordsworth, Coleridge, Keats and Shelley), led a major reversal of outlook. The wilderness now stood for order and freedom; the city as a chaotic jungle ruled by social outcasts. Paul Johnson's (1993) *Intellectuals* brands both Rousseau and Shelley as complete scoundrels, who merely justified their own abysmal personal conduct in the name of fidelity to nature. Yet they had great influence. Somewhat later, the founding of the great US national parks, such as Yellowstone (1872) and Adirondack (1885), showed strikingly the revaluation of nature and the wilderness. (Are places like these nowadays the last arena left for displaying old qualities of toughness and virility?) Reversals of perspective are confusing: not only literature but also painting can be studied as a reflection of historical attitudes to the natural environment (see Chapter 2 of Ittelson *et al.*, 1974).

ATTITUDES

I have said that *attitudes* are a form of evaluation. What is meant by this term? The Social Psychology textbook of Baron & Byrne (1991) gives the following definition:

> a relatively lasting cluster of feelings, beliefs and behaviour tendencies directed towards specific persons, groups, ideas or objects.

Let me note one aspect of this definition. An attitude is something intermediate in terms of duration between, on the one hand, a transient *opinion*, and, on the other, a deep-seated *character trait*.

One theory about social attitudes is very well known. It is the theory of "cognitive dissonance" proposed by the American Leon Festinger. Festinger made his career and reputation with this theory and then moved out of Social Psychology altogether and studied visual perception. He says he was running out of fresh ideas on cognitive dissonance, which makes me a bit suspicious of the hundreds of studies that researchers have published on it, but let me describe it. The basic insight is that our attitudes derive from our experience, which is complex and which, when we reflect on it, often reveals contradictions.

An example of cognitive dissonance would be when someone with strong "Green" views drives a petrol-hungry Mercedes in to work each day. He could reduce this dissonance by *adding consonant elements*, for example, recalling that his new car has far better mileage than his old one. Another way to reduce it would be by *minimising the importance of the cognitive elements involved*, for example, thinking that he is only one of millions of commuters. Alternatively, he could *change one dissonant element*, for example, by taking public transport, or by going anti-environmental. Here, he might well choose the path of least resistance, that is, change the cognitive element least resistant to alteration. This last alternative is a major way attitudes *change*.

Sometimes, we act in a manner discrepant with our beliefs. Having poor justification for this creates large dissonance, and so a large amount of attitude change. An experiment by Festinger & Carlsmith (1959) showed this. Subjects did some dull tasks, and were then given monetary rewards for telling other subjects that they had been interesting. Subjects given $1 started thinking the tasks really were interesting, unlike subjects given $20. This "less justification leads to more attitude change" effect only works where the subject has chosen for himself to do the attitude-discrepant behaviour, feels responsible for it and thinks it important.

Unfulfilled expectancies also create dissonance and attitude change (Aronson & Mills, 1959). Certain closed groups "initiate" new members: they may be required to make vows, pay money, perform various wacky assignments, and so on. In an experiment by Aronson & Mills (1959), subjects were initiated into one of several groups, with the initiation varying in severity. Subjects who had undergone a severe initiation subsequently rated much more highly a discussion they took part in within the group. This shows how anything we have worked for, e.g., a qualification, we start to value even more once we have it.

Now, having covered some main issues on cognitive dissonance, I want to take it as established that attitudes are inherently prone to change, and move on.

How are our attitudes formed during our development? Psychologists have isolated three principles of learning that can be applied to the learning of attitudes as well as of many other things. One of these principles is called "classical conditioning". It refers to what happens when a neutral stimulus is regularly paired with a stimulus that elicits a certain response naturally. This "unconditioned response" will transfer to the previously neutral stimulus, and is then called a "conditioned response". Those who know some Psychology will think of this as "Pavlov's salivating dogs".

Everything else being equal, if we have a splitting headache when we meet a stranger, we tend to form a negative attitude to him. If it is a lovely spring day, we are more likely to view him positively from then on. A second principle of learning is called "instrumental conditioning". It refers to the way a response that is rewarded is strengthened and one that is punished is weakened. For example if someone criticises us for holding a certain attitude we become less likely to hold it. The third principle is "social learning", sometimes called "observational learning", or "modelling". This is fairly self-explanatory. Suppose our parents express an attitude verbally—for example, of dislike of the prime minister—or they might show their attitude through their actions, for example, by attending a political rally—then we tend to acquire the same attitude. Of course it is not only our parents who influence us in this way. We can also form attitudes by social learning from our friends, colleagues, and teachers.

With all these three principles, a very important factor is what has been called the Law of Primacy, that *first* impressions are what count. This is not only a matter of what happens when we meet someone for the first time. It would tend to suggest that our mother's opinions are more influential than our father's, and so on. It accounts for the tremendous importance that all social reformers place on education, and the heat about attempts to remove racist or sexist imagery from very young children's books, for example.

So attitudes arise from classical or instrumental conditioning and social learning. Thus an ecological attitude, such as one to pesticides, might be formed by multiple influences: from social learning (seeing our parents use them), from classical conditioning (we once spilled a pesticide on ourselves and became ill), and from instrumental conditioning (the year we did not use one and lost our garden crop as a result).

A major part of the programme of explaining how attitudes develop is explaining also the fact that they tend to cluster together. For example, Brown (1963) describes a cluster of attitudes, which he says go together to some extent at least, and define the political Right. They include loyalty to a father-figure or leader, a high regard for the idea of the family, positive attitude towards discipline, anti-feminism, patriotism, valuing of tradition, religiousness, high regard for property and tendency to prejudice. There is much interesting empirical work trying to check out the extent to which these various attitudes *do* go together.

So there is plenty of basic work to provide a foundation for understanding attitudes to the environment. It is already clear that there are going to be major differences between people with regard to specific issues, which cannot be explained in any simple way. What follows is an attempt to give some flavour of what research work on environmental evaluation consists in. This chapter continues by considering two areas of the field where I have done some empirical work myself. First, whether there are differences in ecological attitudes between people in town and in country, and, second, the importance of familiarity of an environment, with particular reference to landscapes.

ECOLOGICAL ATTITUDES IN TOWN AND COUNTRY

We saw from the definition of "attitude" that it has three components: affect towards the object of the attitude, cognition about it, and behavioural commitment to some kind of action. From the beginning, the possibility of measuring attitudes and using them to predict behaviour has appealed to the applied psychologist. It could not be long before the great surge of interest in the interaction between the physical environment and people led to the construction of attitude scales in this area.

Maloney *et al.* (1975) developed a 45-item Ecological Attitude test instrument that comprised three sub-scales of Affect, Cognition and Commitment. The test obeys standard methodological rules: for example, they took the precaution to mix items from the different sub-scales through the questionnaire. This is useful in view of the known tendency of subjects to perseverate (repeat the same response on successive items). Since their article, the increasing public awareness and alarm about environmental issues has led other investigators to develop other scales. For example Weigel & Weigel (1978) have brought general attitude-measurement theory to bear on the construction of an Environmental Concern scale. Van Liere & Dunlap (1981) have questioned whether their measurement instrument is an important improvement. Also, Albrecht *et al.* (1982) have presented a New Environmental Paradigm scale for measuring environmental concern. They have examined its reliability, validity and dimensionality, in a more general and methodologically sophisticated way than its predecessors. Nevertheless, the Maloney *et al.* (1975) questionnaire has been used the most extensively, has three sub-scales of intuitive validity, and includes questions suitable for a young sample. So Rosaleen McCrorie and I used it in research to be described.

Not all the items of their questionnaire are as appropriate in a European and a child context, as in the American adult one where Maloney *et al.* (1975) standardised it. My collaborator in this research, Rosaleen McCrorie, chose a 20-item subset for this work in Northern Ireland. The population of Northern Ireland is one and a half million spread through six counties but heavily concentrated in the Greater Belfast region. Beyond the simple aim of developing norms for work in a different part of the world, this study sought for possible influences on ecological attitudes of urban/rural background and of gender. Pilot work and consultation with teachers confirmed the level of the questionnaire was appropriate to a young sample. Its broad rationale is given by Maloney & Ward's (1973) depiction of the ecological crisis, and attack on those content merely to theorise about it (p584):

> The solution to the ecological crisis does not lie in traditional technological approaches but rather in the alteration of human behavior. In short, the ecological crisis is a crisis of maladaptive behavior . . . we must go to the people in an attempt to understand these behaviors. We must determine what the population knows regarding ecology, the environment and pollution; how they feel about it; what commitments they are willing to make; and what commitments they do make. These are the necessary antecedent steps that must be made before an attempt can be made to modify critically maladaptive behavior.

There is a long-standing debate about the relationship of attitudes to the behaviour they logically imply. With regard to the environment specifically, one might speculate to what extent people interested in groups like Greenpeace and Friends of the Earth actually take environmental action (see Hummell, 1977). Positive behaviours are not necessarily allied to deep environmental concern. For example, a motorist who pays for a pollution control device on his or her car may be making nothing but a token gesture (O'Riordan, 1976).

Such evidence as there is suggests that, despite the many initiatives on environmental education that have been mounted in recent years, public environmental knowledge remains "painfully low" (Arcury & Johnson, 1987). That study confirmed that the major correlates of environmental awareness are education, income and sex (another is occupation). It also stressed the need for developing environmental knowledge scales that can be used across a variety of people. Age is another important predictor. The Jesuit maxim was that one who is yours at 7 years is yours for life. By questioning young children, and thus partially eliminating the effects of income and education, it is possible to appraise the effect of some other variables. It is also possible to study a relatively high average level of environmental concern. For it is well documented that there is an inverse relation between age and environmental concern, due partly to a secular trend as well as to a genuine effect of ageing (Honnold, 1984). Further support that younger age groups are the critical ones to explore is given by Jaus (1984). He found that a mere 2 hours of environmental education in the third grade had a positive effect (as compared with a control group) on environmental attitudes 2 years later. This is consistent with the finding of Blum (1987)'s cross-cultural study that such environmental knowledge as ninth/tenth graders possess comes from the mass media rather than from school.

McCrorie tested 50 boys and 50 girls from a rural (co-educational) school, and the same number from an urban one in Belfast (Williams & McCrorie, 1990). These were high-ability grammar schools. In other words most children had passed the Transfer Examination taken at age 11 years. The average age of the children was 11.7 years, and did not vary greatly as all were in the same grade.

The probe-statements relating to Affect about the environment were:

(1) People worry too much about additives in food.
(2) It worries me to think that much of the food we eat is treated with chemicals.
(3) It makes me angry to think that the government does not do enough about pollution.
(4) "The world will be dead in 40 years time if we do not do enough about our environment". This statement does not bother me.
(5) I become angry when I think about the harm being done to plant and animal life by pollution.
(6) I am not bothered by loud noise.
(7) When I see the results of pollution I get angry.

The probe-statements relating to Cognition about the environment were:

(1) I would be willing to ride a bike or take a bus in order to reduce air pollution.
(2) I would not join a club concerned solely with environmental problems.
(3) I would be willing to use an alternative means of transport in order to do something about pollution.
(4) It is the government's job to improve things not mine.
(5) I would give money to help improve the environment.
(6) I would stop buying something made by a company that causes pollution.
(7) I would be prepared to give out leaflets or literature about environmental problems.

The probe-statements relating to Commitment regarding the environment were:

(1) I have never bought something simply because it caused less pollution.
(2) I would never write to an M.P. about environmental problems.
(3) I am prepared to do something positive about environmental problems.
(4) I like to buy goods that are packed in containers that can be re-used.
(5) I would go to a meeting to help improve the environment.
(6) I have never taken part in an anti-litter campaign.

What were the results of the survey?

Affect did not correlate significantly with Commitment (although Cognition did correlate about 0.3 with both the others). It seems that this sample is already aware of the trade-off between environmental despoliation and economic growth, of which I have already spoken. Ramsey & Rickson (1976) show it to be important in generating inconsistency between environmental affect and commitment. Another explanation for the lack of correlation is a sense of powerlessness and depression. We need a similar explanation for the finding that people of low socio-economic status, who suffer the effects of air pollution most, do less to campaign against it (Swan, 1970).

When the distribution of scores is bimodal (showing two separate peaks), it suggests the area is particularly contentious. We found no real sign of bimodality for any of the three sub-scales.

Rural children showed significantly higher whole-scale scores than urban children. This finding does not seem to have been anticipated by other literature. Rural children have a greater affinity with the natural environment, know more about its behaviours, and, at least if female, are involved in more ecological concerns. Comparing males with females, the profile on the three sub-scales differed significantly across sexes, in that females tended to be higher on Affect and Cognition, but lower on Commitment. This difference in profile was far more marked in the urban sample than in the rural sample. There was no overall difference between male and female children.

One slightly puzzling finding was that rural males scored low on Commitment. Perhaps it is because they are "tied to the land", and so, possibly, not consciously aware of it. For example, just knowing the difference between types of fertiliser, and the inherent advantages and disadvantages of their use, caught them in a dilemma answering the question about this. For it was they who, in adult life, would, in practice, have to take decisions about it.

However, the sex difference in profile across the scales matches very well the work of Gifford *et al.* (1982). They, too, say that while females express greater Affect about the environment, males have more Commitment. On the other hand they find that males have more knowledge about the environment – in contrast to the pattern here. It should be noted that their subjects were university students, rather than 11-year-olds, which could explain this discrepancy. The finding that the sex difference is significantly more marked in an urban sample extends their work. Gifford *et al.* (1982) discuss their sex differences in terms of the differential socialisation of women stemming largely from sex-role stereotyping.

The correlation between Cognition and Commitment was especially true in rural children. It seems that rural subjects, being involved with the environment, know about the problems, and know what they are doing is right. It is not that they are giving the "socially desirable" response for the sake of appearances. This is analogous to the report of Gifford *et al.* (1982) that environmental education students not only know more about the environment but report more commitment than do other students.

Cone & Hays (1980) argued, with evidence, that behaviour can be modified to produce environmentally constructive outcomes. For example, air pollution can be reduced by encouraging car-pooling. However, there is no evidence that changing one behaviour of someone pro-ecologically will result in a general pro-ecological change. It is necessary to provide individuals with comprehensive information and education regarding environmental issues, and the effects of their behaviour upon the environment. This will form a more environmentally aware individual, and one who is prepared to modify his entire behaviour in a pro-environmental direction.

Clearly, while those from a rural background may be more aware of issues like the environment-friendliness of different fertilisers, those from an urban background may have more negative experiences of littering and graffiti. In many countries now, programmes of environmental education are being initiated. There is even a specialist *Journal of Environmental Education*. I think such programmes can alter people's attitudes for life. The number of primary age children who are vegetarian, no matter what their parents' taste, strikes me. What this research seems to strengthen is the case for the tailoring of environmental education programmes according to the environmental background of pupils (Terence Lee has done recent research coming to a similar conclusion). This means the broadcast media, however benign and powerful their effect in this direction, should never replace programmes customised for the individual school. In this way we can look forward to a different and better environment and world.

In terms of the wider theme of this book, what this research shows is that people differ greatly in their attitudes to the environment. I feel this could only surprise an extreme "nomothetic" or "psychonomic" psychologist, but perhaps the average person also thinks in this way: that people are much the same. However, these differences can be predicted to some extent from knowledge of someone's "group" attributes; here, their urban/rural residence and their gender. In Chapter 4, I shall consider the relevance of these to mental health.

FAMILIARITY OF AN ENVIRONMENT

One attribute of an environment that affects its evaluation is its familiarity. It is an aspect of the detrimental environment that may not strike everyone at first. Most people would recognise, on reflection, the deterrent effect of a strange environment. At the opposite extreme, Bertrand Russell once said that boredom is a much more potent factor in life than people generally recognise.

It was Zajonc (1968), who revived interest in the effects of repeated exposure on liking. In everyday life, we often take for granted the positive value of exposure. For example, companies repeat their television commercials (and advertisements generally) frequently, in the expectation that this will change our attitude to a particular brand. The popularity of "soaps" also may have a lot to do with their frequent repetition. In elections, the candidates tipped to win are generally celebrities, or incumbents, or people who have received much media exposure.

Zajonc took this assumption into the laboratory and found that it was upheld. His experiments used as stimuli Chinese ideographs, "Turkish-like words", and photographs of faces. The "mere exposure" effect has since been found with several other stimuli including the clothes we wear (De Long & Saluoso-Deonier, 1983). Further research has shown that the presence of the effect does not even depend upon recognition of the stimuli (Seamon *et al.* 1983; Wilson, 1979). For repeated stimuli that the subject cannot identify as familiar (e.g., from the neglected ear in "dichotic listening") still attract positive feelings. So it cannot just be that we have a negative "orientation reaction" to some new or low-frequency stimulus that "habituates" (wears off) with repetition. The mere exposure effect also occurs when subjects believe (falsely) that they are seeing all stimuli equally often, as shown by Jorgensen & Cervone (1978). In this study, they deceived the subjects into thinking that they were being shown the infrequent stimulus again. They were told these repetitions were "sub-threshold" (so faint the stimulus was escaping notice), and that this was happening often enough to make all stimuli equally frequent.

"Mere exposure" cannot be the whole story, however, as other investigators have found a *negative* relationship between exposure and evaluation (Cantor, 1968; five studies of Crandall reviewed by Harrison, 1977). The normal positive effect seems to disappear with *simple* stimuli (Berlyne, 1971: 191; Harrison, 1977; Smith & Dorfman, 1975). Such stimuli are liable, of course, to be familiar already. The apparent contradiction can be reconciled by postulating that at some level of familiarity the preference function reaches a turning point and diminishes. In other words, there is an inverted-U relationship between familiarity and liking. As familiarity increases, liking does *not* always increase. This is what it means to say that the graph of the function is "non-monotonic". Sluckin *et al.* (1982) found exactly this function for a variety of stimuli such as letters, words and names. The material point about these stimuli is that they occur naturally in everyday life: their degree of familiarity is a given and cannot be manipulated significantly. Sluckin *et al.* (1982) measured familiarity as a subjective rating rather than objectively manipulating it by repetition.

The rise and fall of records within the pop-music charts is further naturalistic evidence of an inverted-U preference function. Promoters all too aware of the mere exposure effect will move heaven and earth to gain "needle time" for their songs on radio. Safe enough the record will climb the charts, but as sure as night follows day the consumer will eventually have had his fill of it and it will start a slow descent.

I have previously mentioned a form of inverted-U relationship, known as the Yerkes–Dodson Law, linking performance to arousal. It is old work, and in the field of aesthetics, specifically, too, this idea goes back at least to Wundt. The "Wundt curve" related the pleasantness of a stimulus to its *intensity* (e.g., the brightness of a light) non-monotonically. More recently, Berlyne has emphasised a non-monotonic relationship to novelty in his aesthetic theory. But the idea of zero novelty is problematic, and later investigators have inverted the x-axis to interpret it in terms of familiarity or time. Recent research has extended the idea that people prefer intermediates—the "golden mean" of moderation in all things—for several different variables. Thus interpersonal distance can be not only too near, as implied by the "personal space" idea, but also too far (Thompson *et al.*, 1979)—this is why the Arab diplomat is doing the chasing.

According to Harrison (1977: 43), "Zajonc and his associates have argued . . . there is no such thing as 'overexposure'". Indeed, over-exposure is diffcult to find with some naturally occurring stimuli (Colman *et al.*, 1981; Hargreaves *et al.*, 1983). Some stimuli such as first names show the monotonic mere exposure effect in spite of being sampled over a wide or complete range. The explanation given for this is a "preference-feedback hypothesis": once a first name becomes over-exposed (or starts to become over-exposed), parents choose it less, so the name averts the declining part of the function. This accounts for a common phenomenon that has been well established by research: cyclical vogues for first names, and for other cultural items. Such feedback is not possible for stimuli not under voluntary control, such as surnames, letters, and words.

Zajonc does seem, subsequently, to have changed his mind about overexposure. Zajonc *et al.* (1972) found an inverted-U relationship, most clearly with paintings as stimuli. This bears out Harrison's (1977) claim that investigations involving visual designs and works of art are special exceptions to the familiarity-leads-to-liking rule. As the repetition effect has become so much a part of psychological orthodoxy, clearly its stimulus-dependence would justify further exploration.

There are important social implications. Where there is intergroup conflict, many hold that reconciliation will follow more contact (McWhirter, 1983, discusses the case of Northern Ireland). It is natural to cite the Zajonc work to support this. So I have done experimental work on this myself. With synthetic speech I found an inverted-U relationship (Williams, 1987a, see also Inomata, 1983). I argue that the use of computer control to make a "cleaner" experiment is what causes the extension into the declining part of the function. On the other hand, my work confirmed stimulus-dependence: with computer graphics, I found no effect of repetition at all, either positive or negative (Williams, 1987b). This is something I simply cannot explain. I would conclude that at least sometimes intermediate familiarity has the best effects, but this rule will not apply to all environments.

EFFECT OF FAMILIARITY OF A LANDSCAPE

So a robust phenomenon in Experimental Social Psychology is that mere exposure to a stimulus can increase liking for it. This has found a more ecological echo in some work with landscapes, finding that their familiarity predicts preference for them. A landscape is more like a macro-environment than the isolated stimuli I have been discussing so far, and I want to cover this work, too, in some depth (Fig. 3.3).

"Familiarity" can be a similar construct to "identifiability", though sometimes it can mean mere *resemblance* to what we know. Identifiability forms part of the comprehensive landscape preference model put forward by

Figure 3.3 The beauty of landscape

S. Kaplan and R. Kaplan (1975). This variable is just one element in their framework, which they base on an abstract informational analysis. Yet *place* seems a very natural category to which to attribute familiarity. Moreover, it is very easy and natural to see an affective bond between people and places; this is what Tuan (1974) has called "topophilia" and written a book about. The temporal aspect of this bond is that you have to stay a few days in a place to get to like it. This clearly has practical relevance to the Social Psychology of tourist behaviour.

Another of the predictive variables for landscape preference that the Kaplans have proposed during their research is complexity. In Japan, the cultural ideal for a garden is something that strikes the Westerner as very austere. Laboratory studies, too, confirm that reducing the level of complexity, as well as increasing it, can promote liking, and so this variable does not make linear predictions. These experiments used nonsense forms. Like the work I have described on repetition, they prepare us for the possibility that familiarity also may have a non-linear influence upon landscape preference, though intuitions may differ here.

We saw that there seems to be some validity in the old saying that "familiarity breeds contempt". Reflecting on this, it seems to me that verbal labelling of a stimulus is an important development in the growth of familiarity with it. This view accords with visual-hemifield evidence that we perceive familiar material better on the right (contralateral and so neurologically closer to the verbal hemisphere) and unfamiliar material on the left. It also seems to me that the moment of verbal labelling may signal a point at which the preference function turns down. The Kaplans themselves, in their 1982 book, move towards a non-monotonic view of the familiarity–preference relationship. Their new view, too, is based on theoretical rather than empirical work. They believe that, while, on the one hand, familiarity will enhance what they call the "making-sense" component of landscape perception, it will, on the other hand, tend to undermine "involvement".

There has been some empirical work, though. With an American sample, Wellman & Buhyoff (1980) tried to separate the effect of familiarity with a region of the country from the effect of familiarity with the specific landscape scenes shown. They found a positive effect of specific but no effect of regional familiarity. With urban scenes, Herzog *et al.* (1976) found a correlation of about a half between preference and familiarity. However, they analysed further, and factor-analysed their scenes into five groups. What this showed favoured the non-monotonic idea, for they found a *negative* correlation within the most familiar group of scenes. Hammitt (1983) did a research study on a bog that reclamationists had turned into a local recreational environment. He found that visitors' experience of this bog recreational environment increased preference for it. This was true when "experience" included previous visits as well as current-visit experience, though it seems obvious that repeat visitors would show greater preference anyway. These three studies are part of a burgeoning corpus of quantitative work on the prediction of landscape preference, and show what can be done.

I pursued the Herzog *et al.* suggestion of non-monotonicity in a study of my own (Williams, 1985), which used colour photographs of landscapes for preference rankings. Compared with using real landscapes, this technique is convenient. It does have the drawback of over-emphasising cues such as texture, which is the main way we perceive depth in a two-dimensional image, where the normal "binocular" disparity cue is not available. I used two sets each of nine photos, one set of local scenes, one of foreign scenes. Each subject ranked photographs, both for preference and for familiarity (a "within-subjects" design). There was a strong negative correlation between the two for local scenes, and a small but significant positive correlation for foreign scenes. Preference rankings were highly reliable at re-test. Of course, we know other variables affect landscape preference: such as, presence of water or trees, topography, evidence of man's influence, seasonality, foreground vegetation or cloud formations. I checked that familiarity of the scenes did not correlate with any of these, so the finding is not an artefact. Florence Hamill repeated this finding with more sets of photographs and more subjects in an undergraduate project, finding a shift from a strong positive to a strong negative correlation in the same way. The statistical results were even more compelling and establish that a curved non-monotonic function is readily found where stimuli of the right degree of familiarity are used. This brings macroscopic landscape evaluation into line with the other more atomistic forms of environmental evaluation, previously discussed, with respect to the effect of familiarity.

The finding that non-monotonicity is the form of the influence of familiarity upon preference for landscapes is parallel to the finding of Hull & Buhyoff (1983). They used distance rather than familiarity as the "independent" (predictive) variable. But the use of over-simplified scientific methods to establish a clear-cut relationship between two variables in this area should not be allowed to obscure its essential richness and profundity.

The western tradition has created a heritage of images, which determine the way in which we view landscape. This heritage has, until recently, come mainly from painting, and many would flatly deny the relevance of work with photographs. Landscapes evoke deep emotions, and strong attitudes towards conservation (a cause to a large extent presupposing the importance of familiarity as a positive attribute).

Some would fnd a landform/landuse approach to the prediction of preference more useful. This approach would emphasise the great land symbols: mountain, sun, lake, river, valley, field, tree, waterfall, cloud, wave, rock, wind, rain, corn and fruit. It might decry anyone who did not recognise that "Love of the mountains is virtue, love of the sea is wisdom" (Bertrand Russell (1946)) (Fig. 3.4). Yet, Environmental Psychology does objectively suggest that these land symbols are not the only salient variables in the prediction of preference for landscapes.

For some, the variable "familiarity" will show insufficient deference to the history of a landscape. They will have in mind the age-old process of clearance, boundary marking, enclosure and settlement.

Others, who love light and nature, have felt the need to emphasise the dynamic force of landscapes, their unity, their harmony, and other attributes such as vitality, vividness and variety.

Appleton (1975) put forward yet a further view,

> aesthetic satisfaction, experienced in the contemplation of landscapes, stems from the immediate perception of landscape features which, in their shapes, colours, spatial arrangements and other visual attributes, act as sign-stimuli indicative of environmental conditions favourable to survival.

In other words, the modern tourist has not shaken off his long evolutionary history, through which the struggle with the natural environment was everything. (For a readable recent restatement of this sort of idea, see Jared Diamond's *The Rise and Fall of the Third Chimpanzee*, 1991.) From this

Figure 3.4 Love of the mountains is virtue

follows Appleton's well-known "prospect–refuge" theory—that we admire landscapes where we can see without being seen.

Appleton sees himself in an intellectual tradition that takes in Burke. The symbolism of the hazard, the need to place oneself at the mercy of nature and savour its danger, the age-old wilderness, these primordial evocations are part of his theory. It is partly the timeless quality of landscape, the "old man river" that was here before me and will be here after me, that gives it a central place in art criticism. And, too, in psychotherapy, where the enduring can lift us out of our busy worries.

There is much more to landscape evaluation, truly, than science says or probably ever will say. A question of practical significance to the increasingly important tourist industry (Fig. 3.5), though, is how familiarity affects the travel bug, and now we have some empirical evidence bearing on this question. A limited success is a platform for bolder investigation of the psychological influence of macro-environments. In particular, it is important to have a better idea of the influence of the urban landscape,

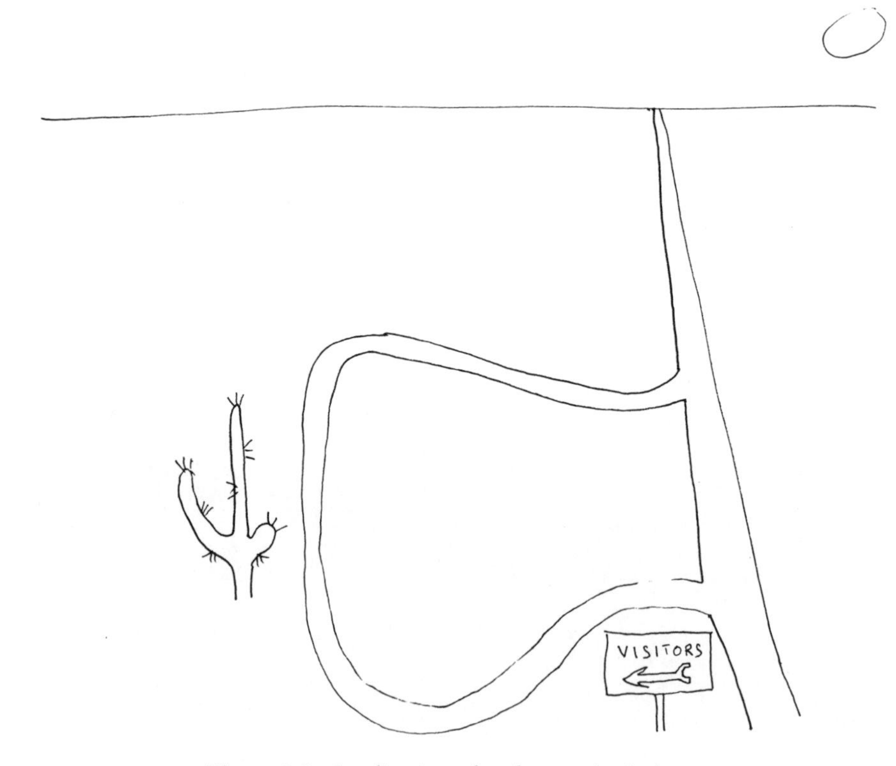

Figure 3.5 Implications for the tourist industry

which may be detrimental to mental health. The very modest aim of this chapter has been to show that "there was nothing to explain it" is an inappropriate response to a mental disturbance. Falling back on brain mythology is a counsel of despair. The environment is unimaginably complex, and our understanding of it, rudimentary in the extreme. With time, if the will is there, we shall understand much better how the environment causes mental disturbance.

ANNOTATED CHAPTER BIBLIOGRAPHY

Because so much of this bibliography is references to empirical research studies, I have made notes on fewer of the items than in earlier chapters.

Albrecht D, Bultena G, Holberg E & Nowak P (1982). The New Environmental Paradigm: measuring environmental concern. *Journal of Environmental Education 13,* 39–43.

Appleton J (1975). *The Experience of Landscape.* London: John Wiley.
A geographer who turned his attention to aesthetics.

Arcury TA & Johnson TP (1987). Public environmental knowledge: a statewide survey. *Journal of Environmental Education 18,* 31–37.

Aronson E & Mills J (1959).The effect of severity of initiation on liking for a group. *Journal of Abnormal & Social Psychology 12,* 16–27.
Aronson wrote a very successful textbook of Social Psychology called *The Social Animal.*

Baron RA & Byrne D (1991). *Social Psychology: Understanding Human Interaction, 6th edn.* London: Allyn & Bacon.
I have been using this book, for teaching the area, since the second edition in 1977. Many students have given me very positive feedback on it. Now 700 pages, with an array of Study Guide, Instructor's Manual and Test Item Bank, it is a complete and very accomplished introduction to Social Psychology.

Berlyne DE (1971). *Aesthetics and Psychobiology.* New York: Appleton–Century–Crofts.

Blum A (1987). Students' knowledge and beliefs concerning environmental issues in four countries. *Journal of Environmental Education 18,* 7–13.

Brown JAC (1963). *Techniques of Persuasion.* Harmondsworth: Penguin.
Like other Psychology books by this author, this book shows a remarkable skill in making complex ideas readable.

Cantor GN (1968). Children's "like–dislike" ratings of familiarized and unfamiliarized visual stimuli. *Journal of Experimental Child Psychology 6,* 651–657.

Colman AM, Sluckin W & Hargreaves DJ (1981). The effect of familiarity on preferences for surnames. *British Journal of Psychology 72,* 363–369.

Cone JD & Hayes SC (1980). *Environmental Problems/Behavioral Solutions.* Monterey CA: Brook Cole.

De Long MR & Saluoso-Deonier C (1983). Effect of redundancy on female observers' visual responses to clothing. *Perceptual & Motor Skills 57,* 243–246.

Diamond J (1991). *The Rise and Fall of the Third Chimpanzee.* London: Radius.
The *third* chimpanzee (you hadn't forgotten the pygmy chimpanzee!) is Man, of course.

Festinger L & Carlsmith JM (1959). Cognitive consequences of forced compliance. *Journal of Abnormal & Social Psychology 58,* 203–210.

Fisher JD, Bell PA & Baum AS (1984). *Environmental Psychology, 2nd edn.* Eastbourne: WB Saunders.
Gifford R, Hay R & Boros K (1982). Individual differences in environmental attitudes. *Journal of Environmental Education 14,* 19–23.
Gould P & White R (1974). *Mental Maps.* Harmondsworth: Pelican.
An appetising taste of behavioural geography, illustrated, of course, with many interesting maps.
Hammitt WE (1983). The familiarity–preference component of on-site recreational experiences. *Leisure Science 4,* 177–193.
Hargreaves DJ, Colman AM & Sluckin W (1983). The attractiveness of names. *Human Relations 36,* 393–401.
Covers everything the parent-to-be, among others, could want to know about the theory of this; a very readable review.
Harrison AA (1977). Mere exposure. In L Berkowitz, *Advances in Experimental Social Psychology.* New York: Academic Press.
An excellent review of the research following up Zajonc (1968).
Herzog TR, Kaplan S & Kaplan R (1976). The prediction of preference for familiar urban places. *Environment & Behavior 8,* 627–645.
Honnold JA (1984). Age and environmental concern: some specification of effects. *Journal of Environmental Education 16,* 4–9.
Hull RB & Buhyoff GJ (1983). Distance and scenic beauty: a nonmonotonic relationship. *Environment & Behavior 15,* 77–92.
Hummell CF (1977). The effects of induced cognitive sets in viewing air pollution scenes. Unpublished doctoral dissertation, Colorado State University.
Inomata S (1983). Research on repeated exposure effects of the stimulus. *Japanese Journal of Experimental Social Psychology 23,* 39–52.
Ittelson WH, Proshansky HM, Rivlin LG & Winkel GH (1974). *An Introduction to Environmental Psychology.* New York: Holt, Rinehart & Winston.
A classic textbook: though I should say "Introduction" was a bit of a misnomer at the time. There is a collection of background readings in Proshansky *et al.* (1978).
Jaus HH (1984). The development and retention of environmental attitudes in elementary school children. *Journal of Environmental Education 15,* 33–36.
Johnson P (1993). *Intellectuals.* London: Orion.
Johnson had a very public conversion from 1960s socialism to 1980s Thatcherism. Surely not all intellectuals can be as bad as he paints this set?
Jorgensen BW & Cervone JS (1978). Affect enhancement in the pseudorecognition paradigm. *Personality & Social Psychology Bulletin 4,* 285–288.
Kaplan R (1975). Some methods and strategies in the prediction of preference. In EH Zube, J Brush & R Fabos, *Landscape Assessment.* Stroudsburg, PA: Dowden Hutchinson & Ross, pp 118–129.
Kaplan S (1975). An informal model for the prediction of preference. In EH Zube, J Brush & R Fabos, *Landscape Assessment.* Stroudsburg, PA: Dowden Hutchinson & Ross, pp 92–107.
Kaplan S & Kaplan R (1982). *Cognition and Environment.* New York: Praeger.
A major work covering much more than landscape preference. A bit abstract for my taste.
Kaplan S, Kaplan R & Wendt JS (1972). Rated preference and complexity for natural and urban visual material. *Perception & Psychophysics 12,* 334–356.
Maloney MP & Ward MO (1973). Ecology: let's hear from the people. An objective scale for the measurement of ecological attitudes and knowledge. *American Psychologist 28,* 583–585.

A journal that can afford to be very choosy.
Maloney MP, Ward MO & Braucht CN (1975). A revised scale for the measurement of ecological attitudes and knowledge. *American Psychologist 30, 787–790.*
McWhirter L (1983). Contact and conflict: the question of integrated education. *Irish Journal of Psychology 6,* 13–27.
The author is a leading authority on the psychological effects of the Troubles on the people of Northern Ireland.
O'Riordan T (1976). Attitudes, behavior and environmental policy issues. In I Altman & JF Wohlwill, *Human Behavior and Environment: Advances in Theory and Research, Vol 1.* New York: Plenum.
Proshansky HM, Ittelson WH & Rivlin LG (1978). *Environment Psychology: Man and his Physical Environment.* New York: Holt, Rinehart & Winston.
A collection of background readings to Ittelson *et al.* (1974).
Ramsey CE & Rickson RE (1976). Environmental knowledge and attitudes. *Journal of Environmental Education 8,* 10–18.
Russell B (1946). *History of Western Philosophy.* London: George Allen & Unwin.
Seamon JG, Brody N & Kauff DM (1983). Affective discrimination of stimuli that are not recognised. II. Effect of delay between study and test. *Bulletin of the Psychonomic Society 21,* 187–189.
Sluckin W, Hargreaves DJ & Colman AM (1982). Some experimental studies of familiarity and liking. *Bulletin of the British Psychological Society 35,* 189–194.
This journal is very widely held in Britain. The article is a brief, readable review of several experiments of the authors.
Smith DF & Dorfman DD (1975). The effect of stimulus uncertainty on the relationship between frequency of exposure and liking. *Journal of Personality & Social Psychology 31,* 150–155.
Swan JA (1970). Response to air pollution: a study of attitudes and coping strategies of high school youths. *Environment & Behavior 2,* 127–152.
Thompson DE, Aiello JR & Epstein EM (1979). Interpersonal distance preference. *Journal of Nonverbal Behaviour 4,* 113–118.
Tuan YF (1974). *Topophilia: a Study of Environmental Perception, Attitudes and Values.* Englewood Cliffs, NJ: Prentice-Hall.
Not a run-of-the-mill textbook.
Van Liere KD & Dunlap RE (1981). Environmental concern: does it make a difference how it is measured? *Environment & Behavior 13,* 651–676.
Weigel R & Weigel J (1978). Environmental concern: the development of a measure. *Environment & Behavior 10,* 3–16.
Wellman JD & Buhyoff GJ (1980). Effects of regional familiarity on landscape preferences. *Journal of Environmental Management 11,* 105–110.
Williams SM (1985). How the familiarity of a landscape affects appreciation of it. *Journal of Environment Management 21,* 63–67.
The photography is my wife's.
Williams SM (1987a). Repeated exposure and the attractiveness of synthetic speech: an inverted-U relationship. *Current Psychological Research & Reviews 6,* 148–154.
Williams SM (1987b). Repeated exposure to computer graphics images: the disappearance of a relationship. *Social & Behavioural Sciences Documents 16,* 52, MS No 7220.
My most important finding in this area, but because it is a "negative result" it has to be buried in this esoteric journal.
Williams SM & McCrorie R (1990). Ecological attitudes in town and country. *Journal of Environmental Management 31,* 157–162.

Wilson WR (1979) Feeling more than we can know: exposure effects without learning. *Journal of Personality & Social Psychology 37,* 811–821.

Zajonc RB (1968). Attitudinal effects of mere exposure. *Journal of Personality & Social Psychology Monograph Supplement 9 (No 2, Part 2).*

The author is a neuropsychologist who has moved into Social Psychology. In my eyes, everything he touches turns to gold.

Zajonc RB, Shaver P, Tavris C & Kreveld DV (1972). Exposure, satiation and stimulus discriminability. *Journal of Personality & Social Psychology 21,* 270–280.

Zube EH (1974). Cross-disciplinary and intermode agreement on the description and evaluation of landscape resources. *Environment & Behavior 6,* 69–89.

CHAPTER 4

Environmental Genesis of Mental Illness

I am but mad north-north-west; when the wind is southerly, I know a hawk from a handsaw.

W. Shakespeare, *Hamlet*

WATCH YOUR LANGUAGE

The way biological assumptions currently tend to go by default is shown quite acutely by the way they are built into the very language we use. "Mental illness" is a metaphorical use of language. Typically, it is *bodies* that are ill, with tangible malfunctions of physical systems. It is an extension of the term "illness" to apply it to minds.

It is a natural enough extension. Suppose you are a general practitioner attacked by a woman patient of thirty-four who refuses to accept that her human immunodeficiency virus test is negative. Anyone could agree that it would be tempting for you to say "this woman is mentally ill". Yet it *is* an extension.

The extension of "illness" beyond simply what goes wrong with bodies presupposes that mental illness is a biological malfunction. Putatively, since people often claim something like "the brain is the organ of the mind", mental illness is something that goes wrong with the brain. People commonly say that mental disturbance is a biochemical imbalance in the brain. The purpose of this book is to propose explanations in terms of the environment, in terms of the way we are made rather than the way we are born. So it is necessary to begin by revising terms.

We need neutral terms that leave open the possible truth of, but do not presuppose, organic explanations of the mental states that trouble people. I propose the three "D"s, mental disorder, disturbance or distress, to be used to cover the same domain of behaviour or mind that people currently call "mental illness". These terms are less familiar than "mental illness", and it is easy to slip into talking of mental illness all the same. This should be unobjectionable if you understand it as a metaphor or "creative collocation", but it is in terms of the "D"s, and particularly mental disturbance, that I myself wish to be understood. I think that many sufferers from mental disturbance will be ashamed and frightened to be thought ill. Though conversely it is probably true, that it relieves some to have an apparently simple explanation of their distress, lifting personal responsibility for it from their own shoulders.

Associated with the term "mental illness" is a whole repertory of other medical terms: "diagnosis" of such diseases as schizophrenia, manic–depressive psychosis and neurosis, each with a certain "prognosis". Underlying the terminology is the reality of the heavy involvement of the medical profession in the "treatment" or care of mental disturbance. And undeniably, they bear most of the final legal responsibility to the community, and to the sufferer, for the outcome of the disturbance. Many doctors would accept the assertion of Foster (1992: 399) that "schizophrenia is a very imprecise diagnosis, rather similar to one of cancer, or heart disease".

In this context, it has to be pointed out that a major alternative terminology comes from Sociology, and particularly from the labelling theory of Thomas Scheff. He regards such terms as "schizophrenia" as labels rather than diagnoses. They are labels none the less with enormously potent consequences for those labelled, and that therefore come to cause terror in those who believe they have been so labelled.

There is certainly a question mark over psychiatric diagnosis ever since Rosenhan's (1975) celebrated study. In this, he told normal volunteers to present themselves to psychiatric hospitals with the single symptom "I hear a voice saying 'Thud' " and thereafter to act totally normally. The hospitals detained every volunteer in mental hospital for several weeks and discharged them with the diagnosis "schizophrenia in remission". On the other hand, is not diagnosis for acknowledged physical illness sometimes pretty unreliable as well?

Psychiatrists draw a basic distinction between disturbances of mood (mania and depression) and disturbances of thought. Unfortunately the distinction does not always seem so easy to draw in practice. One early study, by Sir Aubrey Lewis, re-examined a group diagnosed as having manic–depressive psychosis and labelled half actually schizophrenic. A review of the question of the reliability of psychiatric diagnosis can be found in Clare's (1976: 125–135) *Psychiatry in Dissent*. Even the definition of someone who is psychotic (as opposed to a neurotic), as having lost touch with reality, may not always be so easy to judge in practice. Also, many feel that diagnoses like alcoholism, personality disorder, drug addiction, anxiety state and post-traumatic stress rarely fit precisely the particular individual they meet on the ward.

The view of mental disturbance as a disorder of brain biochemistry is part of what is often called, usually critically, the "medical model". Many mental health professionals are perfectly happy with this model, but those generally called, collectively, "antipsychiatrists" have sharply attacked it.

This group believes the disturbance is caused by alienation (being "turned off") in an inadequate society (Fig. 4.1). They believe the disturbed are often less of a menace than the supposedly healthy. R.D. Laing (1967), whose name I shall be repeating, said that, "The perfectly adjusted bomber pilot may be a greater threat to species survival than the hospitalized schizophrenic deluded that the Bomb is inside him". They emphasise that like all previous societies, ours is a very imperfect one (so, there have been well over 100 million war deaths this century). The mental services user, who does not adjust to it, may be the one who is right. Mental disturbance can even be the price some people pay for their creative achievements, for it is not hard to find examples of creative geniuses who had episodes of madness. Antipsychiatrists also would say words like "schizophrenia" are simply a form of character assassination, used to invalidate and attach a lasting stigma to people and to what they say, in a twentieth-century form of witchcraft. It seems clear enough that for many years in the Soviet

Figure 4.1 Is he the victim of an imperfect society? Reproduced by permission of Mirror Syndication International

Union, the old tyrannical society abused psychiatry. The medical model has been compared to a television viewer sending for the TV repairer because he dislikes the programme he sees on the screen (Szasz, 1973).

Some people find the whole tenor of this woundingly unappreciative of what mental health professionals do. One psychiatrist has said, "The Establishment psychiatrist has become a kind of bogeyman for the underground, like the policeman. This is a paranoid idea" (Boyers and Orrill, 1972). How can anyone regard the Yorkshire Ripper, say, as a victim of an imperfect society? How can a conscientious doctor withhold medication from someone who might otherwise commit suicide?

Why is the language so important? Well, people use the medical model to justify some dramatic physical interventions. There is "psychosurgery" (i.e., irreversible neurosurgery for a psychological disturbance), and electro-convulsive therapy. Treatment with neuroleptic drugs also can have a serious neurological side-effect, called "tardive dyskinesia". I should acknowledge, however, that practical necessity may make neuroleptic treatment the choice even for someone whose disturbance is environmental in origin. The environment may lie in the individual's past, or in an enduring structure of society, neither of which can be easily remedied directly.

Those scarred by physical treatments may become fair game for any quackery less noxious to the sufferer. I do not know the full extent of this problem, but suspect it to be great. A touchingly gentler physical intervention, is a combination of penicillin and evening primrose oil, reported by Vaddadi to have a dramatic positive effect on some schizophrenic patients (Foster, 1992: 429).

The antipsychiatrists tend to hold that the disturbed behaviour is a natural form of healing that should not be "short-circuited" by means of drugs. The terminology does matter. This sort of problem could affect *you*. "Admission to hospital for in-patient treatment" is the fate at some time or other in their lives of well over one in ten of the population. Actual admission is just the tip of the iceberg. A quarter of all women experience a major depression (Charney & Weissman, 1988) and this figure may be going up.

"SCHIZOPHRENIA"

This is a specific example of charged language, modern jargon for "madness". More than any other epithet, we use "mad" to dismiss and deny the normal rights of someone else. When Ronald Reagan called the Libyan leader Colonel Gaddafi "a mad dog", it was the prelude to a terrible punishment.

A recent book by Irving Gottesman (1992), *Schizophrenia Genesis*, sets out the medical version in this area. It gives the lifetime prevalence of schizophrenia as one in a hundred adults (p38). These people have twice the mortality rate of the general population, one in ten of them committing suicide (ten times the normal rate). They also have fewer children than normal, because fewer of them marry (p196). Laing said that no-one can understand schizophrenia who has not understood despair (though I doubt that Gottesman has much time for Laing).

Notice, however, the prevalence rate is for adults. Doctors do not call psychoses beginning before puberty schizophrenia and so, since the proportion of the population who are pre-pubertal varies greatly across countries, quite technical Weinberg "age-correcting" is necessary to estimate prevalence. Misdiagnosis of schizophrenia in people who have suffered head injury, or even just the flu, are no longer accepted as valid. A diagnosis of "reactive schizophrenia" may well catch on, when doctors hear that the geneticist Gottesman (1992) found that it did *not* carry any increased risk for mental illness among the offspring of the "reactive schizophrenic" (p102). The diagnosis of schizophrenia seems to have "arrived" at the beginning of the nineteenth century. Then, it "seems to have become visible all over the Western world and to have increased rapidly for a hundred years" (p2). Now, we find an 80% decrease in the mental hospital patient population in the Western world over the period 1955–86. One study mentioned by Gottesman found secular changes in "fashion" in diagnosing schizophrenia in the US over the period 1925–70. Another study, of contemporary usage, showed that "broad" diagnostic criteria yield five times as many schizophrenics as do "narrow" ones. Yet Gottesman can make the strange assertion that "It seems safe to conclude that the incidence of schizophrenia in most human populations round the world is rather similar" (p80). Gottesman states that individuals can be diagnosed as schizophrenic even if they are currently in remission and have few or no symptoms currently. He goes on, "This position is unfair to individual patients and could be misused against their civil rights, but it is necessary to advance the scientific quest for the origins of the disorder" (p36). I envy him his confidence

For there is no doubt that for many people today the diagnosis carries a horrific stigma. The media take up every incident of violence, especially sexual assault, by anyone with this diagnosis, in a way that paints a totally false impression, as Gottesman agrees (1992: pp190, 194). I rather like the idea that he puts forward of changing "schizophrenia" to the "Pinel–Haslam syndrome". But it is far too late for Hitler's psychiatric victims—80 000 approved for compulsory sterilisation before the war, and then 100 000 "mercy deaths" during it, under a programme headed by Hitler's associate Philip Bouhler.

In the medical and scientific literature, the assumption that schizophrenia is a brain disease is rife. Purely from the pattern of deficits, and without so much as an internal brain scan, Opler *et al.* (1991) conclude that schizophrenia results from prefrontal dysfunction.

There is a review of possible neurological aetiologies of schizophrenia by Berquier & Ashton (1991).

LIFE STRESS AS THE CAUSAL FACTOR

What is the cause of mental disturbance: is it inborn or is it made by the environment, whether current environment or developmental history? Let us deal with the simpler cases first.

There are occasions where the causes of mental disturbance are organic: head injury, encephalitis, epilepsy, cerebral tumours, dementia (whether of old age or earlier), Korsakoff's syndrome, cerebral arteriosclerosis, "organic neurasthenia", and others.

Conversely, there are cases where attempts to correct supposed faulty learning histories by means of "behaviour therapy" are very successful, at least in eliminating the obvious symptom. So that environmental causes for the problem seem very likely. I am referring mainly to the phobias (from fear of spiders to the disabling "agoraphobic" fear of going outdoors) and obsessive–compulsive disorder.

It is over the huge middle ground that the debate rages. An environmentalist would naturally put someone's mental disturbance down to stress in his or her life, to "problems in living" such as physical illness, bereavement, divorce, unemployment and the like. In the research literature, life stress has been the object of increasing attention in modern times. It is worth pondering—why did it take so long? Connolly (1985) wrote:

> Some psychiatrists find themselves a little impatient with [this] research effort. They are quite used to making sense, with their patients, of life's happenings and seeing them as causal to the conditions presented for treatment. This is reflected in the nomenclature. We talk of a reactive depression or an hysterical reaction.

The environmentalist stance can be taken to the point of claiming that anyone will crack under pressure. It is likely that few, if any, normal stable adults can withstand a determined and prolonged attempt at brainwashing. This view gained fresh credence as elite Allied pilots were paraded on Iraqi television relayed to the rest of the world, as apparently "different people", during the war to liberate Kuwait. This seems some justification outside the field of orthodox psychiatry for an idea of reactivity. It is worth noting that the work that more than any other put Stress on the map as an explanatory construct was Selye's (1978) idea of a General Adaptation syndrome. This was originally observed as what people called "shell shock", during the First World War.

On the other hand, this sort of research is bedevilled by a vagueness of definition of what is genuine life stress. Many readers will have encountered pointcounts for life event checklists (100 stress points for moving house, 200 for a bereavement, and so on)—a landmark in such thinking was a paper by Holmes & Rahe (1967). No doubt they would be the first to acknowledge its original crudity. Monroe & Simons (1991) give references to more recent developments, which also list material on chronic stressors and daily hassles. There could hardly seem a better example of life stress than the "Troubles" of Northern Ireland. Yet researchers often claim that they have had little psychological effect—in some bad areas, *decreases* in depression have occurred, as though this sort of "excitement" can actually lift the spirits. If one adopts a dogmatic environmentalism, it is easy to slip into regarding any and every major life event as a stress. Remember the view of "crisis" built into the Chinese ideograph for that word, that it is risk plus opportunity.

The many dozens of studies of the effects of life stress upon mental disturbance do consistently point to the existence of a definite though variable association, a correlation of about 0.3 (Cochrane, 1983: 115). But is the association so variable simply because it is difficult to be certain with any real person just when the disturbance actually began? Or is it because it is difficult to know what is a life stress for a particular individual? One recent study found that, for some, joining the Army is enough to induce schizophrenia-type symptoms.

The stress must precede the disturbance for a causal hypothesis to be secure, but since the disturbance is itself likely to produce life stresses, it is easy to understand skepticism about causality. For it can hardly be denied that irritability, or fatigue, or lack of concentration, or social withdrawal, for example, may cause such life stress events as divorce, or loss of a job. Even if one accepts that the stress does precede the disturbance it could be argued that the cause of the disturbance is still biological, the stress merely precipitates or "triggers" it.

I am going to talk about terminology again here. As a proponent of environmentalism myself, I none the less feel comfortable with the idea of biological dispositions. A biologically given temperament (Gordon Claridge (1985) has advanced this notion persuasively in this context) may underlie a disturbance that stress precipitates. But the word "trigger" suggests it is the disposition that is all important (the "gun" ready to go off at the squeeze of a finger), whereas I would hold it is the precipitant that is important and about which something can be done to aid the sufferer. The word "diathesis" is sometimes used now to refer to the biological disposition to be contrasted with stress as a causal factor. It has been claimed that interaction between the two causes is easier to conceptualise with this terminology. Though it has been said that it is meaningless to assign weights to the two sorts of cause, I should say that it does matter if we are to orient ourselves towards prevention rather than cure. Prevention is neglected, because its success is so rarely recognised and acknowledged.

Orthodox psychiatry draws a basic distinction between disorders of mood, such as depression and mania, and disorders of thinking, such as delusions and hallucinations – often dubbed "symptoms of schizophrenia". It is for depression that the best evidence of a causal factor for stress exists. The well-known studies of Brown & Harris (see their 1978 book) carefully showed (to the satisfaction of many) that stresses do have a causal rather than merely a "trigger" role in the onset of disturbance. Eight or nine out of ten diagnosed depressions were preceded by stressful life events in the previous 9 months. This compared with only three out of ten without depression experiencing such stress. Many other investigators have now reported similar findings. How about economic stresses specifically? Intuitively, I should expect that unemployment, which singles some people out, would have a more disturbing effect than inflation, which the whole community shares. This is indeed one finding of such "life stress" research. Though another explanation is that recent unemployment is also a form of loss, which Brown & Harris show to be the main type of event associated with depression.

Another line of thought supporting the life-stress model is the psychopathological potency of disasters. By following up the victims of disasters one can usually evade the objection to stress theory that people create their own stresses. An example that comes to my mind immediately was a report of an explosion at sea, where no less than three-quarters of the crew of hundreds subsequently suffered psychiatric disorder. But there have been many formal studies.

The terrorist pub bombings in Birmingham in 1974 not only killed 21, but sent 159 to hospital, 116 of whom were allowed home later in the evening. A 2-year follow-up of these 116 showed that they suffered prolonged "symptomatic disability" and disturbance in social relationships: 70% described prolonged emotional symptoms; 40% had required psychotropic medication and some had become dependent upon it. Several people who had been moderate to heavy drinkers before their experience had subsequently shown an alcoholic drinking pattern (Sims *et al.*, 1979).

There was another study (Ganzini *et al.*, 1990) in the USA, of 72 people who lost all their retirement savings in a banking collapse. Twenty-one of them suffered major depression during the following 2 years (though only four of these sought help from a mental health professional).

Indeed, the American Psychiatric Association's Diagnostic and Statistical Manual Version III explicitly recognises a category of "post-traumatic stress disorder". A survey of 1000 people in their twenties (Breslau *et al.*,

1991) found 394 reporting exposure at sometime in their lives to traumatic events and 93 of these having post-traumatic stress disorder.

Losing someone to whom one is attached

Bereavement, as well as less final separations, commonly causes depression. Depression is a disturbance of mood, feeling sad, heavy, lethargic, the opposite pole to manic excitement. But it also can carry with it a cognitive disturbance.

The depressed person takes an irrationally negative view of the world, seeing every dark aspect and none of the good ones. Even his dreams, as recorded by psychoanalysts, are more like nightmares. Above all he has a black picture of himself, with feelings of worthlessness (absence of self-esteem, severe self-criticism), and feelings of despair and hopelessness about his future. Very often, and strangely to the observer, the bereaved have feelings of guilt about the death. It seems that, if possible, he will blame himself for it in some complicated way, or if not, reproach himself for doing or saying the wrong thing the last time he saw the departed. More straightforwardly, he may experience strong regrets about what might have been, joint projects left undone, love left unexpressed.

Freud had an odd but intriguing theory that the bereaved has a pool of unconscious hostility towards the deceased loved one, perhaps blaming her for having left him. Because he knows this hostility to be irrational, and cannot allow himself to express or even experience it, he directs the anger against himself. More prosaically, the death may occur at the end of a long illness with much suffering. It is natural and common for the bereaved to feel some release and liberation when the death finally occurs, but it is about these feelings that he may feel guilt.

Experimental psychological studies have also shown the depressed person to be more likely to identify with "losers" who have been rejected or punished. Further, they are unnecessarily pessimistic about their chances of completing a demanding task. However, if they succeed in such a task, they do better than average on a second one.

There is also a self-feeding process in which the bereaved recognises that he is sad, assumes that he will always be sad, and so feels even sadder. He also recognises his own apathy, which saddens him further, and so things snowball. It is likely that the ability to mourn well helps us to age well and accept the coming of our own death. One idea, is that mourning is most successful when the death is allowed to cause the minimum disruption to the flow of everyday life. It is normal, to see the mourner pass through stages of shock, denial and numbness. It may be better to leave him alone through this process. It also may help to accept or even, with sensitivity, encourage a strong display of emotions, however embarrassing they may be to those around the bereaved. It also may help to assure the mourner that his painful memories will fade or ease, even if they never go completely, and it must be recognised that the period of intense mourning for a loved one will last in the order of years rather than months.

Earlier experiences of loss, playing a role of "traumata", sensitise the mourner further to the current one. These memories, which may have been buried for many years, come back to plague him at this time.

The leading name on the Psychology of attachment and separation is John Bowlby. Beginning his work in mid-century, he maintained the position that "mother love in infancy and childhood is as important for mental health as are vitamins and proteins for physical health". This thesis had enormous influence on the paediatric profession through its comprehensive statement in the three-volume *Attachment and Loss* (Bowlby, 1973). Another psychologist, Erik Erikson, explained that through interaction with its mother the infant achieves, all being well, "basic trust" of the world in general and other people in particular.

If an enforced separation threatens the bond with the mother, for example, when she goes into hospital for another birth, the child is liable to become "clinging", when it regains the mother. But while the bond is right, the child can feel secure in moving further away from the mother, emotionally as well as in the physical sense. The mother is a "base" from which the child can achieve vital developmental tasks of exploring strange or merely unfamiliar environments.

Freud held that it is because the mother satisfies its primary drive for food, that the infant develops the attachment. So, it is a secondary drive, a theory commonly known as "cupboard love". Studies of animals have moved psychologists away from this theory. Thus, Harlow (1971) found that baby monkeys preferred a cuddly adult to one that provided food. Lorenz (1981) found goslings "imprinting" on their mothers just because she was around at a "critical period" in the gosling's life.

One issue that remains unresolved is whether the infant is programmed by instinct to attach itself to a single individual, an idea known as "monotropy".

Antidepressant drugs, the tricyclics and the mono-amine oxidase inhibitors and the seroton in reuptake agents, play a major role in treatment of depression. But they have the unfortunate effect of telling the bereaved that his disturbance is not a natural result of his loss, but rather reflects an imbalanced body chemistry.

Malady of the oppressed?

Among chronic life stresses prevalent in the community, the one that stands out is of course poverty; the environmentalist position naturally predicts that this will be associated with mental disturbance. A typical early study found the lowest social class, five, accounted for twice as many first-admission mental patients as expected by chance. It would take too long to explain all the possible mechanisms for this. An interesting recent variant is that poor people's diet (in particular lack of essential fatty acids, leading to prostaglandin imbalances) predispose to schizophrenia (Foster, 1992: 420). There is now little doubt that poverty is associated with schizophrenia. It remains a debatable question whether poverty causes schizophrenia (the classical position of Faris & Dunham), or whether schizophrenia causes poverty (the view of Jarvis)—see Goldstein & Goldstein, 1984.

The latter view is sometimes associated with an ideology that the poverty of schizophrenics is "Nature's way" of preventing them from reproducing themselves. We see the same theme with IQ. It is the constant refrain of the eugenicist Right, that State Income Support encourages people with inferior IQs to have large families, and so is lowering the IQ of future generations. In other words, misplaced compassion is thwarting the selective and hereditary effect of poverty. It is difficult to rehearse this ideology dispassionately and I implore my readers not to take it seriously.

Guilt by association is, of course, a faulty line of argument. It is possible to see common sense in Jarvis's view (without regarding it as by any means the whole story). Gottesman (1992) describes the later research in 1969 by Goldberg & Morrison (1969) that also emphasised downward social drift in the "incipiently schizophrenic". The idea is more colourfully expressed by George Orwell in *Down and Out in Paris and London*,

> The Paris slums are a gathering-place for eccentric people—people who have fallen into solitary, half-mad grooves of life and given up trying to be normal or decent. Poverty frees them from ordinary standards of behaviour, just as money frees people from work.

One aspect of poverty that some have sought to associate with psychological disorder is the likelihood that it compels living in more crowded conditions. So, too many social contacts and too much unpredictability might overload the pauper. The research evidence is unclear (Kellett, 1984) for many reasons. It is difficult to take account of the social skills of the individual. Are markers and barriers available that help to maintain boundaries? What about personality factors, or whether one is close to friends or to strangers? The easiest measure of crowding is number of persons per room, which can be extracted from census forms. But surely it matters what size the room and the household are, and how easy it is to escape from them.

The apparent importance of what one might call "social power", in deciding who actually receives diagnosis, or "labelling", of mental illness, is another reason for taking an environmentalist perspective very seriously. For research consistently reports that a large proportion, perhaps as high as 75%, of those suffering "clinical psychiatric disorders" have never been in treatment. Clearly, some of these people regard it as a stain on their character to receive psychiatric treatment, and have the power to avoid it that the client of the welfare state does not. The Midtown Manhattan study (see Srole *et al.*, 1961) was based on 1600 people, chosen as a random sample of the population of that area. It found 24% to be plainly psychologically impaired, while another 58% had mild to moderate symptoms; only 18% were considered well. This is an urban area that might show a higher incidence of stress and stress-related disorders; but the same picture emerged in the Stirling County study (see Leighton *et al.*, 1963) (sample of 1000) of a rural area: 33% manifested clear disorder, while only 17% were well. The many who are abjuring treatment, that objectively they could do with, are probably more adept at what Laing (1967), in a famous title, called *The Politics of Experience*. Social power seems to decide, not only who receives diagnosis/ labelling, but also the nature of the diagnosis, with the label/diagnosis "schizophrenia" being assigned more often to the powerless. One research study asked psychiatrists to diagnose people from case histories. These were altered, so that two histories varied only in the educational level or occupational status given: a lower position led to a more severe diagnosis (Cochrane, 1983: 162).

THE IMPORTANCE OF SOCIAL SUPPORT

The critical importance of the social support offered to the sufferer in rehabilitating him from mental disturbance also bolsters the environmentalist perspective. Cochrane (1983: 122) leaves us in no doubt that social support is an important factor in recovery from mental disturbance. All my own clinical experience shrieks it out too. That may well apply to many physical illnesses as well, but not so dramatically, and possibly only insofar as there is a psychological component in most physical illness anyway.

Clearly this factor has implications for the major social policy debate over the relative use of "community" as opposed to institutional care. One participant in that debate is Olsen, who, in 1985, concluded that it is possible successfully to accommodate and rehabilitate both the recovered and the chronically mentally disturbed within "substitute family care". Also, this arrangement has many advantages over the more traditional solutions of hospital and family care. Any rationally planned system of mental care will surely weigh this very carefully for the user. What are the relative merits of family, community and hospital care with respect to the support they offer him or her? Anyone who has ever been involved with someone mentally disturbed, will have pondered this choice, and recognised its relevance to his future prospects.

Environmentalists also frequently emphasise the importance of detrimental circumstance within the sufferer's family itself, in the origin and development of his mental disturbance. Thus, Arieti (1974) has taken the strong position, that family maladaptation is *always* present with schizophrenics. Laing, moreover, took the even stronger view that a particular kind of family maladaptation always precedes the onset of schizophrenia (see Laing and Esterson, 1965). This is the kind of family maladaptation known as the "double bind". Parents of the sufferer-to-be will send him conflicting signals simultaneously. For example, a mother might say "John, come and hug mother", in a tone of voice that suggests that she would find nothing more repugnant than that John should hug her. The message is that John should show love to his mother, but also that if he does show love to his mother she will find it repulsive. Faced with an unlivable situation, says Laing, the symptoms of schizophrenia develop in John as an attempt to find a way out of the situation. Broadly, Laing seems to see the whole of schizophrenia as a complete and despairing withdrawal of the sufferer from a family reality that he or she finds utterly intolerable. Clearly, if Laing's view holds any plausibility, it will be a mistake to return remitted schizophrenics to such families unless the families also can be changed.

More recent work in this tradition has focused on "high emotional expression" families. In a study of patients returning to emotionally charged homes, 12 families received an educational programme, and there was only one relapse, compared with six relapses out of 12 controls. This MRC Social Psychiatry Unit study has been confirmed several times (see Kuipers, 1992).

EFFECTS OF INTERVENTION

The view that mental disturbance is analogous to physical illness, with a probable biological and therefore plausibly genetic cause, is partly due to the existence of distinct and distinctive mental institutions. It was Russell Barton (1976) whose name is associated with the view that these institutions cause a special "institutional neurosis". This was responsible, he claimed, for all, rather than merely some, so-called "schizophrenic" symptomatology. In support of this claim something should be pointed out that many forget or do not know. The "open door" policy of easier discharge from mental hospitals dramatically improved the general quality of life in such places even before the arrival of neuroleptic drugs. Yet the latter are so widely credited with the transformation of the prognosis for those mentally disturbed.

Apart from their dreadful neuromuscular side-effects, it seems likely that the phenothiazines cause some of the excess mortality in diagnosed schizophrenics. They do this by increasing sedentary habits and weight and so increasing deaths from heart disease.

The perhaps mistaken emphasis upon the importance of neuroleptics is part of an emphasis upon physical treatment that tends naturally to go with a biological and hereditarian orientation. That these medicaments do have an importance can hardly be denied. Yet the view of life-stress workers Brown and Harris is that: "attention to a person's environment may turn out to be at least as effective as physical treatment".

Almost by definition, madness is behaviour that we cannot understand. So, there is a strong temptation to assume that its causation is also incomprehensible or mysterious, e.g., bound up with the poorly understood biology of the brain. *A priori*, then, comes this assumption of physical causation. Yet Gillie (1976) says that: "research into the biochemistry of [schizophrenia] has a long history of failure relieved only by what seems to be never-ending optimism". Thus, a contemporary example of such research is that on opioid peptides. An exhaustive review (Jeffcoate, 1985) of much expensive research on these can conclude,

> There is at present little firm evidence to support a role for endogenous opioid peptides in the pathogenesis of psychiatric diseases. Moreover attempts to treat such diseases with opiate antagonists and opioid peptides have been largely unsuccessful.

Other biochemical avenues may turn out to be more promising. Never the less, perhaps it is not mental sufferers alone, who can learn from what is said to have occurred to one of the greatest of all psychologists, William James (see Schultz and Schultz, 1987). He recovered from a severe neurasthenic ("nerve-weakness") episode mainly by dint of *releasing* himself from the notion that all mental disorder required to have a physical basis. According to Foster (1992: 404) there have been historical changes in the relative emphasis on biochemistry, with this approach currently resurgent after a long period in which a psychosocial approach to schizophrenia was dominant.

THE PHYSICAL ENVIRONMENT

I emphasised, in Chapters 2 and 3, the importance of our better understanding of the psychological effects of the physical environment. What implications does this have for mental disturbance?

First, what are the effects of urban living? It is clear that hospital admissions show a higher rate in urban areas—they are highest for Britain in Greater London. This could be because urban dwellers live closer to hospital, or simply cannot find enough space in the family or community, or are less sensitive to the stigma of admission (Cochrane, 1983: 62). Yet it is unlikely that there is not also an underlying non-artefactual difference in mental disturbance itself, though Freeman (1984: 12) concludes that it must be small. Freeman does think (p13) that there may be an optimum size beyond which the costs of cities start to outweigh their benefits, resulting in an escalation of levels of social pathology.

Similarly Jonathan Freedman (1975) has argued that crowding does not necessarily have detrimental effects but simply intensifies whatever effects are happening anyway. My own view is that such positions are unlikely to represent the whole of the truth. For example, it is widely acknowledged that there is a tendency towards severe psychological problems in those who move from small town or rural environment to inner city (Cochrane, 1983: 64).

It is, after all, widely believed that living in cities is bad for one's physical health as well: in the graphic simile, living in New York is equivalent to smoking 38 cigarettes a day. However, in this respect, too, people have denied a worse standing for cities. Thus, Winsborough (1965) used a statistical technique called "partialling out" to eliminate the effects of social class and some other variables. He found that the density (of population) relation with death rate, often reported, was reversed, i.e. the lower death rate was associated with higher density living conditions (Cochrane, 1983: 66). (This is a major research issue and not at the heart of Psychology. But one contrary recent study, found that the pace of life in cities corresponds strongly with the incidence of coronary heart disease. The measure of pace of life was a better predictor than the personality trait, through which it is supposed to be mediated, the "Type A" lifestyle of all rush and bustle.)

On a more specific issue, it seems to be widely held that rehousing, particularly when compulsion is applied insensitively, is a factor in causing mental disturbance (Fried, 1963). Three million people were so displaced in Britain in the period 1955–75. There is a phrase "New Town blues" to refer particularly to the problems of women with small children at home. Notwithstanding this, there is little evidence that mental disturbance shows a higher rate in New Towns or housing estates than in the typical suburb. Though researchers often say that migration, whether international, inter-regional (especially from country to town) or within a city, can increase the risk of mental disorder. Giggs (1984: 351) concludes any effect is complex and will be partly a consequence of disturbed people migrating in greater numbers than others (partly to escape stigma) anyway.

The fantastically high costs of development land in the city were one reason for the popularity of the high-rise idea for residential as well as office use. Such tower blocks, as they are rather chillingly called, have often been criticised in the media for social problems they are claimed to induce, as well as on aesthetic grounds. The writer Paul Johnson said, in 1991, "there has been an architectural disaster of unparalleled proportions [in Britain], which has no precedent in our history and the dimension of which we are only just beginning to grasp". On the other hand it has been said that there are only two good studies actually demonstrating negative effects of living in tower blocks (Cochrane, 1983: 76). This source takes the view that tower blocks are simply an example of an environment against which prejudice has grown up. However, there is evidence that in multi-storey flats disturbance is more common among occupants of higher floors, particularly in high-rise blocks. It seems that single people generally like high-rise living, as do many childless couples, but families do not (Ineichen, 1979).

There is clearly a question of to what extent social attitudes to design are the result of irrational factors (there is a book by Stuart Sutherland (1992), emphasising the prevalence of *Irrationality*). One major factor is certainly traditionalism, shown, above all, in the continuing popularity of the classical style. There are the "pediments", and four styles of column (Doric, Ionic, Corinthian and Tuscan), seen throughout the country, from the British Museum to new housing estates. It has been said that nineteenth-century Englishmen turned to Gothic and stone for their new Houses of Parliament "to express the continuities of an ancient constitution". More likely it was simply the style of the day. Currently we use the phrase "the heritage industry" as implicit criticism of this facet of British life.

A further aspect of the city as an environment that should be mentioned is greater noise level. Early in the days of the car many thought it would not last because of the intolerable noise nuisance it posed. What is the relevance to psychiatric morbidity? Tarnopolsky and Clark (1984: 261) took the imaginative step of investigating residents affected by the Heathrow flight path, but found no evidence of an effect. They emphasise annoyance as the main social and medical problem produced by noise. In line with this, the striking phrase used by the British Noise Abatement Society is "this form of assault".

One other stressor is climate, which I discussed in the previous chapter. Many people do not realise that there is a temperature difference between town and country (which crowding exaggerates).

In the end though, can one really make a general evaluation of the city as psychologically detrimental? About half the world's population live in towns or cities. How can one bracket together say Bath and Birmingham? In published British comparisons of town and country the town is often a Northern one with a bad image (sometimes expressed by the adjective "grimy"). Is any British town comparable to what we generally call "shanty towns" found around cities in the developing world? I think of Rio de Janeiro, or São Paulo, where well over half the inhabitants live like this. In the poor South of the world, the number of urban dwellers has risen fivefold, to one and a half billion, since World War II. Mexico City is now more than twenty million. Very rapid expansion of a great city such as Cairo means that systems designed to cope with a population a tenth the size are being overwhelmed. On the other hand, is rural life such an idyll? Or can it not rather mean isolation, loneliness and lack of stimulus? The melancholy of peasant life is a frequent theme of literature, such as André Gide's *La Symphonie Pastorale*, or the anthropological work *Tristes Tropiques* of Claude Lévi-Strauss (1992). Indeed, it was the Industrial Revolution, and the accompanying cultural movement of Romanticism, that led to a great positive revaluation of country life. Thus, it has been said that "for good Wordsworthians a week in the country is the equivalent of going to church". No doubt, this had something to do with the technological taming of nature, made possible by the earlier agricultural revolution.

There is also the question of regional differences. In Britain one would want to consider the north–south (Smith, 1989), as well as the urban/rural distinction.

This work is in its infancy. It could be a giant that is awakening.

ANNOTATED CHAPTER BIBLIOGRAPHY

Arieti S (1974). (Ed) *Interpretations of Schizophrenia I.* New York: Basic Books

Barton R (1976). *Institutional Neurosis, 3rd edn.* Bristol: John Wright.

Beck AT (1976). *Cognitive Therapy and the Emotional Disorders.* Harmondsworth: Penguin.

The author is well-known for his "cognitive theory of depression".

Berquier A & Ashton R (1991). A selective review of possible neurological etiologies of schizophrenia. *Clinical Psychology Review 11,* 645–661

Bowlby J (1973). *Attachment and Loss. Vol 2 Separation.* New York: Basic Books.

A doctor in the psychoanalytic tradition, who wrote this huge and hugely influential three-volume work. The film "John" (and two others) by his students, James and Joyce Robertson, illustrates his idea about early separation very dramatically.

Boyers R & Orrill R (1972). *Laing and Antipsychiatry.* Harmondsworth: Penguin.

Breslau N, Davis GC, Andreski P & Peterson E (1991). Traumatic events and post-traumatic stress disorder in an urban population of young adults. *Archives of General Psychiatry 48,* 216–222.

Brown GW & Harris T (1978). *Social Origins of Depression.* London: Tavistock.

The classic of "life events" research. Based on comprehensive empirical work in Camberwell, London.

Busfield J (1986). *Managing Madness.* London: Unwin Hyman.

An authoritative and recent text for the whole mental health field.

Charney E & Weissman MM (1988). Epidemiology of depressive illness. In JJ Mann, *Phenomenology of depressive illness.* New York: Human Sciences Press.

Clare A (1976). *Psychiatry in Dissent.* London: Tavistock.

A psychiatrist heavily involved with the mass media, and currently Director of St Patrick's Hospital in Dublin (founded by Dean Jonathan Swift). Stuart Sutherland has recommended this book for its balance.

Claridge GS (1985). *Origins of Mental Illness: Temperament, Deviance and Disorder.* Oxford: Blackwell.

Cochrane R (1983). *The Social Creation of Mental Illness.* London: Longman.
Too good and short not to read. Persuasive because it never shouts.
Connolly J (1985). Life happenings and illness. In RN Gaind, FI Fawzy, BL Hudson & RO Pasnau, *Current Themes in Psychiatry, Vol 4.* London: Macmillan
Ellis A (1987). The impossibility of achieving consistently good mental health. *American Psychologist 47,* 364–375.
Ellis says *no-one* is immune from mental disarray.
Foster, HD (1992). *Health, Disease and the Environment.* London: Belhaven.
Freedman J (1975). *Crowding and Behavior.* San Francisco: Freeman.
High marks.
Freeman H (1984). *Mental Health and the Environment.* London: Churchill Livingstone.
Five hundred pages that you will have to read if you do not believe the argument of this chapter. Freeman, a British psychiatrist, is the leading authority on the research literature described by the title of the book he has edited. The book has 17 chapters and over 1000 references to the primary literature.
Freeman H (1989). Mental Health and the urban environment. In R Krieps, *Environment and Health: A Holistic Approach.* Aldershot: Gower
No advantages over the author's book, except for greater recency.
Fried M (1963). Grieving for a lost home. In LJ Duhl, *The Urban Condition: People and Policy in the Metropolis.* New York: Basic Books.
Ganzini L, McFarland BH & Cutler D (1990). Prevalence of mental disorders after catastrophic financial loss. *Journal of Nervous & Mental Disease 178,* 680–685.
Gide A (1925). *La Symphonie Pastorale.* Paris: Gallimard.
Giggs JA (1984). Residential mobility and mental health. In H Freeman, *Mental Health and the Environment.* London: Churchill Livingstone.
Gillie O (1976). *Who Do You Think You Are?* London: Hart-Davis.
Goldberg EM & Morrison SL (1969). Schizophrenia and social class. *British Journal of Psychiatry 109,* 785–802.
Goldstein M & Goldstein I (1984). *The Experience of Science: An Interdisciplinary Approach.* New York: Plenum.
Gottesman II (1992). *Schizophrenia Genesis.* San Francisco: WH Freeman.
I say more about this book in Chapter 7.
Holmes TH & Rahe RH (1967). The Social Readjustment Rating Scale. *Journal of Psychosomatic Research 11,* 213–218.
The landmark paper for checklists of life events quantified for stressfulness. Often criticised now for oversimplifying—also the events chosen obviously apply to men rather than women.
Ineichen B (1979). High rise living and mental stress. *Biology & Human Affairs 44,* 81–85.
Harlow HF (1971). *Learning to Love.* San Francisco: Albion.
Jeffcoate WJ (1985). Enkephalins, endorphins and psychiatric disease. In RN Gaind, FI Fawzy, BL Hudson & RO Pasnau, *Current Themes in Psychiatry, Vol 4.* London: Macmillan.
Johnstone L (1989). *Users and Abusers of Psychiatry.* London: Routledge.
A former clinical psychologist shows sympathy and understanding for people discarded by some sections of society, with particular insight into the problems of some women (in her Chapter 5). Noteworthy for a telling drive at the giant pharmaceutical companies, who bear a lot of the responsibility for the "medical model".

Kellett JM (1984). Crowding and territoriality: a psychiatric view. In H Freeman, *Mental Health and the Environment*. London: Churchill Livingstone.

Kessler RC, Price RH & Wortman CB (1985). Social factors in psychopathology: stress, social support and coping processes. *Annual Review of Psychology 36*, 531–572.
A scholarly review of the "coal-face" (the primary literature).

Kuipers L (1992). Expressed emotion research in Europe. *British Journal of Clinical Psychology 31(4)*, 429–443.

Laing RD (1967). *The Politics of Experience*. Harmondsworth: Penguin.
And the Bird of Paradise.

Laing RD & Esterson A (1965). *Sanity, Madness and the Family. Vol 1 Families of Schizophrenics*. New York: Basic Books.

Leighton DC, Harding JS, Macklin DB, Hughes CC & Leighton AH (1963). Psychiatric findings of the Stirling County Study. *American Journal of Psychiatry 119*, 1021–1026

Lévi-Strauss C (1992). *Tristes Tropiques*. Harmondsworth: Viking Penguin.

Littlewood J (1992). *Aspects of Grief*. London: Routledge.

Mangen SP (1982). *Sociology and Mental Health: An Introduction for Nurses and Other Care-Givers*. London: Churchill Livingstone.
A straightforward brief textbook.

Lorenz K (1981). *The Foundation of Ethology*. New York: Springer-Verlag.

Masson J (1989). *Against Therapy*. London: Collins.
Radicalism reaches a new frontier: Masson thinks even "talking" therapies often bad in practice for the user. Sensitivity to her environment needs to be heightened much further.

Monroe SM & Simons AD (1991). Diathesis–stress theories in the context of life stress research: implications for the depressive disorders. *Psychological Bulletin 110*, 406–425.
Up to scratch for the journal.

Olsen MR (1985). The care of the chronically mentally ill: boarding out an alternative to family and hospital care. In RN Gaind, FI Fawzy, BL Hudson & RO Pasnau, *Current Themes in Psychiatry, Vol 4*. London: Macmillan

Opler LA, Ramirez PM, Rosenkilde CE & Fiszbein A (1991). Neurocognitive features of chronic schizophrenic inpatients. *Journal of Nervous & Mental Disease 179*, 638–640.

Rosenhan D (1975). On being sane in insane places. In T Scheff, *Labelling Madness*. Englewood Cliffs, NJ: Prentice-Hall.
An astounding study showing that perfectly sane people can be diagnosed schizophrenic. A classic that still discomfits Mental Health Service managers.

Schultz DP & Schultz SE (1987). *A History of Modern Psychology 4/e*. London: Harcourt, Brace, Jovanovitch.

Sedgwick P (1982). *Psychopolitics*. London: Pluto Press.
A difficult condensation, but easier than reading the other difficult or many-volumed authors covered, such as Laing, Szasz and Goffman.

Selye H (1978). *Stress of Life*. New York: McGraw-Hill.

Sims ACP, White AC & Murphy T (1979). Aftermath neurosis: psychological sequelae of the Birmingham bombings in victims not seriously injured. *Medicine, Science & the Law 19*, 78–81.

Smith D (1989). *North and South*. Harmondsworth: Penguin.

Srole L, Langner TS, Michael ST & Opler MK (1961). *Mental Health in the Metropolis*. New York: McGraw-Hill.

Sutherland NS (1976). *Breakdown.* London: Weidenfeld & Nicholson.
An eminent Psychology professor's account of his own period of *Sturm und Drang* occurring at midlife. Fairly sympathetic to the medical model.
Sutherland NS (1992). *Irrationality.* London: Constable.
Szasz T (1973). Mental illness as metaphor. *Nature 242,* 305–307.
A persistent and prolific American critic of conventional psychiatry, best known for his soundbite that "mental illness" is a myth. Szasz was described as a "radical libertarian" by the television psychiatrist Jonathan Miller. Many years ago he founded a journal, *The Abolitionist,* whose platform is to repeal mental health legislation.
Tarnopolsky A & Clark C (1984). Environmental noise and mental health. In H Freeman, *Mental Health and the Environment.* London: Churchill Livingstone.
Wing JK (1978). *Reasoning about Madness.* Oxford: Oxford University Press.
A medically trained researcher defies that any behaviour (so-called "schizophrenia") is truly incomprehensible.
Winsborough HH (1965). Social consequences of high population density. *Law and Contemporary Problems 30,* 120–126.

CHAPTER 5

The Therapeutic Environment for the Mentally Unhealthy

For we were nursed upon the self-same hill.
J. Milton, *Lycidas.*

BACKGROUND

A Hospital Plan based on population estimates and norms for beds decides the extent of hospital provision for those mentally disturbed in a particular Health Authority. For example, in 1989, I was working for the North East Essex Health Authority, which had a population of about 300 000. On this basis, it had (figures include all health services including general medicine) 200 medical posts, 3000 nursing posts and 2000 other staff posts. The beds provided were 600 in mental health and 700 in learning disabilities. Outside these mental services, there were a further 1100 beds. Many users of the mental health services live with their own families, and another important aspect of service is to provide care that gives the family a respite from their own "informal" care. There is also day care, of course, and, as I shall go on to discuss, "community care".

INSTITUTIONAL PSYCHOLOGY

The growing awareness, of the detrimental consequences for care, of providing it in any kind of *institution,* underlines the importance of family care. Once, people held the old asylums in great pride as a symbol of a progressive society's concern for its weaker members. I have an office in the grounds of one (Fig. 5.1). From its window I can see overflowing evidence of the horticultural love that ground staff assisted by patients have lavished on it over the best part of a century. I am luckier than some users, and I think it is important to give thought to windows and views onto the pastoral beauty. Would that be just tinkering with a jalopy that needs to be scrapped and replaced, though?

Goffman's work on "total institutions", showed that "disculturation", or "untraining" occurs, rendering the user unable, at least temporarily, to

Figure 5.1 Severalls Mental Hospital today

manage many features of outside life, if he ever returns to it. The institution undermines the person's sense of self, though unintentionally, through a series of procedures that are degrading and humiliating. This was the focus of a book by Russell Barton on *Institutional Neurosis*, whose first edition came out as long ago as 1962. He argued, at first to deaf ears, that the accepted "symptoms" of "chronic psychosis" were not the effects of a physical disease of the brain. Apathy, submissiveness, greyness, an aura of deterioration, a lack of any initiative or interest in the environment, let alone the future: Barton denied they were a physical illness. Rather, they could be understood in terms of the asylum progressively stripping new admissions of the psychological crutches we all take for granted.

It is easy to forget how much our image, for ourselves as well as for others, depends on our clothes. "We are what we wear" says Bull (1975)—a typical supporting study is that by Ericksen & Sirgy (1989). And again, "A person has three parts: body, mind and clothes" (Bull, 1975). In the asylum, the user wore hospital clothes. Much of our personal budget goes on clothes, which we often discard before wearing them out, partly to keep up with powerful but ever-changing fashions. Most of us care what others think about the way we dress, and take derogatory remarks about our clothes personally. At a party, it is a social disaster if two women turn up in the same dress. Yet the user wore hospital clothes.

Most of us have built up an expectancy that, when we speak, we are fairly likely to be answered. If the user spoke, he was unlikely to be answered. If staff spoke to him, other than in a question, they did not expect a response. Other users did not provide a normal communicative environment. His conversational skills evaporated, and psychotic verbal behaviour might be his only way to gain a reaction from those around him. He was expected to obey, and to conform.

Do we always realise how much our sense of personal competence depends on the regular achievement of small routine tasks? In the asylum, the bed was made for you, the meals were provided, the washing up was done by staff. In time, some users came to believe that they were incapable of doing these things themselves. Discharged from the asylum, probably in penury, they quickly found a way to return to it. Their skills of financial management would have been severely eroded. The outside community now rejected and abused them. Their ability to structure a day of activity for themselves was weakened. They might drift into petty crime. They missed their buddies, and their carers, also those other users who were in an even worse state than themselves.

It is easy to forget how attached we all become to our own territories—a seat in a public place, the area on a beach marked by our towel—until someone challenges us about them. Over many years this "territoriality" may become astonishingly potent. Researchers, first encountering long-stay wards, have used phrases like "petrification" (Holahan, 1979) and "institutional sanctity" (Sommer)(Fig.5.2).

Today, in Britain, Barton's view is an accepted orthodoxy. People looked at the prisons, and saw the same features of institutionalisation, in those whose deviance did not present as mental disturbance. They looked at asylums where the staff were conscious of the institutional problem, so that psychotic phenomena were less frequent. Efforts to maintain some contact with the outside world helped, as did a programme of activities. Care to avoid authoritarianism, and to make allowances for the user's loss of friends and possessions (even birthday cards), paid dividends. Above all, a healthy skepticism about the virtue of the psychotropic drugs made a difference.

Now, the policy of "community care" has drastically reduced the populations of the old asylums.

Change needs to be handled with sensitivity. Machiavelli (1532) was an early analyst of the problems in "instituting a new order of things". Much has been written about the need for someone to act as a "change agent" and the value of this agent being senior within an organisation. It is often difficult to persuade members of the organisation that change is really possible, or desirable.

Paradoxically, those who remain in the asylums now, are those who are

Figure 5.2 Severalls Hospital interior today

the most dependent on them, who may also revel in the freedom from the overcrowding they remember from years ago. Consulting these users, one may find a very positive attitude towards the asylum, and a strong lobby against further changes to it. The end of overcrowding also may mean loss of friends, though, so feelings about the thinning out may be mixed.

The basic problem, of institutional neurosis, remains very much a live issue. This will be apparent to anyone who reads the *Report on Services for Mentally Ill People and Elderly People in the Torbay Health District* (NHS Health Advisory Service and DoH Social Services Inspectorate, 1991). Although it says (p1) that "Torbay is regarded by the DoH as a pioneering Health Authority", the report has hard words for the General Hospital Psychiatric Unit, called the Edith Morgan Centre. It says (p18) that

> In spite of the original design brief, the unit has a surprisingly institutional atmosphere. Some patients and community-based staff have spoken of the pervading atmosphere of boredom and aimlessness.

Later one reads (p29),

> Furniture and fittings are standard. The ward has no warm atmosphere, with little or no evidence of personal belongings in either the ward or the dormitories upstairs. Bathrooms, showers and toilets are adequate but with no homely features.

In my view, this situation can come about partly because so many mental care professionals are what the Canters (1979a) call "hit-and-runners". This includes doctors, social workers, clinical psychologists, speech therapists and physiotherapists at least. Because they spend so little continuous time in the users' physical environment, they do not appreciate its importance. Anyhow, the user is generally summoned to meet them in a private room, and given little opportunity to form a relationship with them.

The motto is "distance lends enchantment". The doctor who holds himself, as some do, remote from his users, enters a self-feeding process. The users read his implicit messages to hold off. They comply, behaving in ways the doctor reads as communicating unwillingness to enter a close relationship. In fact, it may signal no more than deference, suppressed indignation, or, in those with major problems, hopelessness. All the health care jobs are in an era of greater and greater "credentialisation". Letters after your name send the news that you are a professional. Nurses have recently found that they have to work for a diploma as well as satisfying the requirements for registration with the state. The aim is to produce a more cerebral, broadly educated nurse. But where will this leave the many nurses, without diploma or registration, who are probably, in practice, providing most of the hands-on care? Will they be professionals too?

In physics one speaks of "centripetal" forces such as gravity that impel an object towards the centre and "centrifugal" forces such as angular momentum that impel it away from the centre. By analogy the terms "sociopetal" and "sociofugal", first coined by Osmond (1957), are gaining currency. Some physical environments pull people towards each other into close interaction; other environments discourage it. A circular seating arrangement is sociopetal, if the chairs are facing inward, sociofugal, if they are facing outward. Holahan (1979) provides confirmation that seating arrangements do have a major social influence in psychiatric hospitals. Since interpersonal interaction is so basic to a therapeutic process, the ideas are consequential for mental health environments. It is necessary to bear in mind, though, that good interaction may depend on the user having a choice whether to enter into conversation or even into simple proximity. Design arrangements that blatantly promote interaction may be less beneficial than more flexible resources. In particular, I believe a blend of fixed and movable seating is often desirable.

One study described in Lee (1976) found that the proportion of what it called "isolated passive" behaviour in one mental facility increased regularly with room size (and consequently with the number of persons sharing the room). As an intervention, a sun lounge was provided, with extra comfortable and attractive furniture, "laid out in ways conducive to good conversation". One month later, when "behaviour mapping" was repeated, the sun lounge's share of total social activity in public rooms had increased from 25 to 42%. Among the readjustments was a drop in "isolated passive" behaviour in the corridors from 32 to 5%. This is in the same vein as the opinion given by Ittelson *et al.* (1974) that central day rooms should not overwhelm the user with too much space or too many possible two-person relationships. Also, Fisher *et al.* (1984) survey empirical evidence (p307) that the number of beds in a room and the number of users assigned to a day room have discernible effects on behaviour. There is an ironic feature of institutions. Users congregate in corridors, leaving organised social and activity areas under-used. But this is because often the corridor is the place where anything happens.

I have been concentrating upon the physical rather than social environment, in tune with the field of Environmental Psychology. I recognise that the social is far more important, but in the context of therapy the physical has a major advantage. It is relatively easy to gain agreement that the old asylums are unsuitable and something needs to be changed. The issue is less threatening than recognising the need for changes in the social environment. Yet it does have great repercussions on the social. Take one aspect of importance.

An aspect of the social environment that is closely intertwined with the physical is the level of staffing for the building. Very roughly, I would say, the more rooms, the more staff are appropriate. The number of close colleagues they have is very important to staff themselves and to fostering a positive social climate. Naturally it is a, perhaps the, central management issue. Finance is a great constraint. Yet if the community sees clear benefits flowing from a humane staffing policy, tailored to fit the mould of the physical facility, the money will surely be forthcoming.

I believe there is a link between the physical environment and prevailing therapeutic orientations. I cannot help feeling that "behavioural" techniques of therapy seem less inappropriate in the depersonalised physical environment of the mental institution.

Currently, they are common. For example, some people have a great fear of getting dirty, and may feel compelled to wash their hands a hundred times a day, to the point where it severely constricts their lifestyle. The behavioural procedure known as "flooding", forces them to endure very dirty environments, carefully bedecked with house dust, cigarette ash and general grime. They may have to wear clothes that have not been washed in recent memory, and so on. Once the user has given consent to the "treatment", however much he may protest, he is not released until the phobia is beaten. It does work. Yet many recognise, that unless you go to the root of a problem, all that will happen, is that another symptom will substitute itself for the one successfully eliminated. His previous treatment may have so alienated the user that he will not seek further help, and so he will be recorded, erroneously, as a complete "cure". I believe that this is happening a great deal. It surely emphasises the importance of *agreement* on a care plan, with *informed consent* of the user.

Much has been written and said about behaviour therapy. To me it seems that, in it, what Buber (1970) called the I–It (rather than the I–Thou) seems dominant. Behaviourism is the logical terminus of the *scientific* approach in Psychology. It emphasises system, rigour and precision, which are all easier with research on animals. Agreement amongst a scientific community is easier when studying observable behaviour rather than internal processes.

A contrasting approach, of humanism, draws on the humanities (language and literature, history, philosophy, the fine arts, mythology, etc.) for its understanding of the person. It emphasises creativity, lived experience and interpretation. It inclines towards soul or spirit, rather than mind, as the subject matter of Psychology.

A middle way between the two, may be the idea of "social science". Abraham Maslow (1954) and Carl Rogers (1951), both humanists, have dominated the application of Psychology to mental nursing. But they are fringe figures within Psychology. Mainstream Psychology has moved from behaviourism to cognitivism, which studies internal processes but still prefers the methods of experiment. It includes the study of the human processing of information, trying to arrive at "black box" models. Psychoanalytic theory is a proto-cognitivism applied to mental disturbance. It emphasises the way emotions originating in the user's childhood can be *transferred* to the care-giver. In this picture, handling the transference, including weaning the user away from it, is central to the therapeutic process.

CUSTODIALISM

Still today, the custodial element of the function of mental facilities shows itself in older buildings simply by their appearance. "Heavy-duty" is the best word I can find to describe it. The isolated situation of these buildings, too, is important to institutional psychology (though some have been brought to the edge of towns by urban sprawl). I am not speaking here of the custodial needs for mentally disordered offenders, which I know little about. The well-known environmental psychiatrist Julian Leff (1983) has said that the compromise between privacy and easy observation in a mental health unit is an architectural challenge yet unsolved.

According to Ronco (1972), in the past, at least, it was staff needs, rather than user needs, that decided design. It is conceivable, that a ward design that eases staff functioning, might also cause a user to sense an overwhelming loss of personal control and privacy. Such feelings may, in turn, contribute to the user's becoming over-dependent on the hospital and withdrawing from normal activities. Diagnosed schizophrenics, who are desperate to withdraw, often found it difficult, usually in the interests of staff surveillance.

Let us return to pioneering Torbay (NHS, 1991: 20):

> [The] Extra Care Area: within an internal secure area, closed off by a digital lock system, are two four-bedded units and three rooms designed for seclusion. The obvious seclusion rooms (with perspex observation panels and heavy-duty door-locking mechanisms) dominate the Extra Care Area and emphasise the custodial atmosphere. We were informed that between a quarter and a third of all admissions are initially placed in this area; most admissions at night (about two each week) are placed in this area first.

It cannot be ignored, that legal compulsion is used to detain people on grounds of their mental state. There is a good discussion of this by Bean (1980). Some laws reflect social consensus, e.g., law about murder; others reflect the current equilibrium of a social conflict, e.g., trade union legislation. Which type is the law about mental health? The early Mental Health Act of 1890 was very "legalistic", in that it strongly safeguarded against what was, at the period, the lifetime disaster of admission to a mental hospital. The next Mental Health Acts were in 1930, 1959 and 1983. From an early stage, some criticised the creation, through legal involvement, of an association between mental disturbance and crime. In 1923, what was then the "Lunacy Department" transferred from the Home Office (still responsible for criminal justice) to the Ministry of Health. Philosophers of law argued that the state had a right to use law in its role as a "parent" of its citizens. Today, in a sense, the state treats those mentally disturbed worse than criminals. They can be detained without a caution, and with no need for justification, or indeed any public scrutiny of the action. The decision is made, not "beyond reasonable doubt", as in a criminal trial, but merely on "the balance of probabilities". In practice, the Act is used, mainly, for people who deny they are ill (though I have argued already in Chapter 4 such denial is reasonable, at least with regard to the particular word "ill"). The Acts have also given doctors strong protection from litigation by people they have detained. In 1930, the need, at least to "act in good faith and with reasonable care", was removed, and in 1959, the need, even to detain the right patient or provide the right diagnosis. There is plenty of evidence, that the pressure is on the "rule enforcers", to err on the side of caution, by detaining anybody conceivably dangerous or suicidal, i.e., to "keep yourself in the clear". Also the proper administration of the Act is contravened in many admissions, with impatience expressed about "stupid bloody forms". In the USA, four times as many people are admitted to mental hospital compulsorily, as are imprisoned. It would be much too simple to see the successive Acts in Britain as a straightforward progression of more liberal treatment of those mentally disturbed. For example, there has been controversy over a new power, given to mental nurses under the 1983 Act, to hold voluntary patients for up to 6 hours.

POOR DESIGN

The authoritative recent text of Goldberg & Huxley (1992), states (p162) that refining the administrative and architectural requirements, for meeting the needs of those mentally ill, lags behind other areas of progress. It is not so long since conditions were so poor that courts intervened, as in the US case of Ricky Wyatt (1971). The court criticised such basic elements, as good lighting, level of noise, degree of crowding, and cleanliness. Thus, it said that there should be no more than six users sharing a bedroom. Today, many would surely agree that even this is seriously counter-therapeutic. At the State Hospital of Belchertown in Massachusetts, a court directed that a minimum sum should be spent on remedying deficiencies in these areas (see Canter and Canter, 1979a). The American Constitution gives great force to the idea that justice will be done in the end, although to British eyes it seems to create a rather legalistic ethos. In many States of the Union, the law now requires minimum standards of "habitability" of mental facilities to be upheld. I believe this is really more important as a signal, rather than with regard to detailed enforceability. A collection of minor *pros* for a building can outweigh a major *con*, such as an offensive smell from a neighbouring factory. Many small features that excite users and staff will not interest others, and this variability in "attention span" is very important. What is needed, above all, is a publicly visible attitude, that design does matter, and that criticism, though never pleasant, will be heard.

Lee (1976) criticises a mental hospital where users spent most of their time in endless corridors, one of them 3000 feet long, which had monotonous reflecting paint surfaces, and few relieving features. "The walls were also highly sound reflective, causing echoes and strange otherworldly effects." Spivack (1967) documented similar phenomena in American hospitals.

And in Torbay in 1991? (NHS, 1991: 27):

> ECT [electro-convulsive therapy] patients have to be transported from the Edith Morgan Centre to the anaesthetic room of the day surgery operating theatre. This is a considerable journey through a succession of linking corridors and lifts, taking fifteen minutes or even longer when lifts are out of action. The process whereby patients who may occasionally be less than willing to receive ECT, have to be taken with a degree of physical restraint through many areas of the hospital is a cause of distress to all who participate in it or witness it.

Or consider this report in *The Independent* (18 May 1992): there had been a dispersal of users from the psychiatric ward of a general hospital, because of insufficient beds – each user lost their own bed, locker and personal space. The upshot? Three suicides in a year. A spokesperson for the relevant health authority said there was no evidence linking the suicides with the users' new conditions. The question comes back to whether we explain suicide in terms of the user's "diathesis" or in terms of an obviously major stress like this.

The giant mausoleums of the great asylum-building programme stand in stolid indifference to the conclusion of Lee (1976: 73) that users recover faster in small hospitals (see also the book by Canter & Canter, 1979c, especially the chapter by Kenny & Canter). It hardly needs research to recognise that large buildings can induce a sense of personal insignificance. They impart a constant message that feelings of worthlessness, that may have brought someone into treatment, or secondary to being labelled mentally ill, are justified.

Take users with anxiety symptoms, too. It is not hard to find examples of designers' disregard for inherent problems. Thus, they will frequently be unhappy in rooms that have only one exit, or that are disproportionately large, such as institutional-type dining rooms. There are many more specific problems like this. To some extent they will always be with us, while economics dictates that we care for people with very different forms of disturbance in the same place. But when the day comes that the movement (in Britain) to consult users who have not been heard extends to mental users, a great deal more will be learned. For one thing, users dominate some areas of the therapeutic environment, and staff simply do not know what such areas are like to inhabit for long periods. Since, as we have seen, passivity is a major feature of the institutional syndrome, it would be wrong to expect all failings in design to be reported by users. Thus, they may fear nursing staff will take such complaints as personal criticism.

GENERAL HOSPITAL UNITS

Yesterday's solution to the problems of institutional psychology, custodialism and poor design was the general hospital psychiatric unit (often discreetly called Ward *P*). Thus, Stephen Dorrell (the British Minister responsible) wrote the following, in a letter of 1992 to Neil McDougall, a vigorous activist for better design of these units: "Everyone would accept that developing psychiatric units on district general hospital sites has been an advance over inpatient units in remote mental hospitals". Yet he goes on to concede:

> In retrospect, some general hospital units have been too large and impersonal. Their environment has been unnecessarily clinical and, in some, too cramped and claustrophobic. There have also been disadvantages of access associated with district general hospital sites—which have not always been located centrally or have lacked parking facilities for day patients and visitors.

He concludes: "The design of the acute psychiatric unit, on or off a district general hospital site is continuing to be refined. I do not believe we yet have a prototype for the future".

Another strand of thinking sees the move into the general hospital as a wrong direction. The World Health Organisation, as long ago as 1953, stated,

> the more the psychiatric hospital imitates the general hospital, as it presently exists, the less successful it will be in creating the atmosphere it needs; too many psychiatric hospitals give the impression of being an uneasy compromise between a general hospital and a prison.

Continuing in that vein, the Butte County study in the USA concluded, "The great majority of both voluntary and involuntary psychiatric inpatients can be treated effectively in a non-hospital setting" (Lamb, 1979).

COMMUNITY CARE

The direction in which considerable strides have been made, in recent years, is away from the hospital, and into the community, for care of those mentally disturbed. It is in line with the comprehensive anti-institutional philosophy expounded, in many books, by Ivan Illich. It has happened even though, currently, a doctor's legitimacy depends on his association with a hospital.

It has happened in spite of "NIMBY" ("Not In My Back Yard"). There have been many, many controversies aroused by the NIMBY syndrome. The protestation of a building society spokesperson to *The Guardian* writer Melanie Phillips captures and encapsulates something for me. "After all this is an *executive* housing development, not the usual sort of place where you find mental health hostels".

To its credit, this sort of attitude has not deterred Torbay (NHS, 1991: 40): "resources have to be diverted to domiciliary care; central to the Closer to Home project in Totnes was the funding and appointment of a co-ordinator and a volunteer co-ordinator" and (p61) " [this Closer to Home project is one of two examples of] exceptionally good practice in the Authority".

Research is trying to keep up with the move into the community. Thus Leff kept track (*The Independent* 17 July 1992) of users of Friern and Claybury psychiatric hospitals, discharged into staffed group homes of about five people in the community. They showed low rates of vagrancy (only six out of 500), and no signs of deterioration after 1 year. However, after a further year, there were increases in anxiety, phobia, incontinence, psychotic behaviour and friendship loss; but no increase in readmission, or desire for it. A spokesperson for the National Schizophrenia Fellowship pointed out to *The Independent* (17 July 1992) the contrast between these results, on the one hand, and the situation with overcrowded general hospital units, on the other.

One danger, with the policy of community care, is that many institutions are very positive, on balance, and some excellent ones are being closed down. Thus the Henderson Hospital in Surrey, for people with "anti-social personality disorders", is under threat of closure at the time of writing. Their case is that they save far more than the cost of their funding, by reducing crime, with its obvious huge costs for victims and the administration of justice. This is cutting no ice with a new-style "Health Trust" required to make a 6% return on assets, and with no incentive to consider wider social implications.

Also, there is much anecdotal evidence that not all "decarcerations" have been managed as well as Friern/Claybury. I hear myself, and read constantly in the press, that "community care" means homelessness, squalor and destitution, for many former inmates. There are new financial pressures involved in being a Health Trust. These seem to have encouraged some of them to cash in the considerable value of the old asylums in the property market, before making adequate provision for the remaining users. Sometimes they move the users into "bed-and-breakfast" type accommodation, only for them to be driven out by other residents who dislike their behaviour. Applying for a "council flat" (a type of accommodation far harder to come by because of the British government's sell-off policy), may be beyond them, due to the bureaucratic complexity. There are cases of transfer to private residential homes hundreds of miles from their families and familiar environment. Former NHS employees run many of the best of these homes, but there must be a question whether people who "know the system" may use their knowledge to cover over aspects of care. I do not believe that all the long-term institutionalised people who I work with myself can cope in the community without major individualised attention.

One way my local Mental Health Trust has attempted to address these concerns is by providing a complex of four houses, with 15 beds. These give a final stage of rehabilitation for their residents before they return to the community. Some will go into staffed community houses, and others into independent accommodation. These are detached, family-sized houses, situated close together, near the entrance to the main asylum, with an easy walk to shops, post office, day-care facilities and bus stop. Each user has a room of his own. The houses provide an experience of living in proximity to others, with scope for increasing independence, and relearning of daily living skills. Staff encourage residents to take some responsibility for the care of their physical environment, but a housekeeping service is provided. Residents are between 16 and 70 years, of either sex, and have been long-stay or frequent-admission users. The complex is staffed for 24 hours of the day. The use of "nursing models" is eclectic, but the main ones are Orem (1971) and Peplau (1988). Each user has a key worker. This is a nurse training area.

I have analysed some data gathered by the Psychology Service of North Essex Mental Health Services, to yield a result that has what looks like an interesting bearing on community mental health care. These were data on "difficult" users presenting challenging behaviour. Professional staff rated each user on four dimensions: Symptomatology, Social Behaviour, Self-care and Motivation. The result depended on the locality. In the town of Colchester, the Symptomatology ratings were independent of the other three (which inter-correlated among themselves). In Tendring (i.e., mainly the seaside resort of Clacton), it was the Social Behaviour ratings that stood out. Inter-correlations can reflect a genuine overlap of meaning (as with Self-care and Motivation). But there is also a "halo effect", where raters tend to transfer positive or negative ratings from one dimension to another. I think this cognitive error is very common; in these data it never accounted for more than about 10% of the variance in the ratings; none the less it seems an error. This means that the independent ratings were especially accurate, which I should interpret as reflecting their particular importance in that district. In most urban areas, people are probably more tolerant of aberrations of self-care, social behaviour and motivation in their fellow-citizens. So here it would be Symptomatology that would assume particular salience. What is different about Tendring, I should guess, is that there seems more of a deviant subculture there. The important thing for difficult users is whether they manage to join the subculture. A sociologist could be much more specific and concrete about this. The important thing is that communities are various, and so, therefore, is community care.

Community care has not, of course, been a specifically British policy. In France, for example, an interesting experiment, perhaps inspired by earlier experience at Gheel in Belgium, has been the Family Colonies at Dun-sur-Auron and Ainay-le-Chateau. Jodelet (1991) has written a book about the latter, where more than a 1000 male psychiatric hospital users have been placed in nearly 500 different family homes. The hosts are typically smallholders, who take their guests for much-needed extra income. The Colony was founded in 1900, so it is disappointing that there has not been real "incorporation" of the ex-users into the community. There does seem to have been a new recognition that mental users are not wholly alien, expressed, for example, by an increased anxiety of mothers about the psychiatric condition of their own children! The foster families are responsible for ensuring that ex-users take their medication, which must do a great deal to "de-normalise" the relationship.

USER-FRIENDLY DESIGN

Quite old work has investigated the importance of physical design of the therapeutic milieu. Thus, Bayes (1967) considered the influence of colour and form of the environment on treatment of disturbed children. To me, these seem like odd aspects to make a start on. But remember that this was at a time when conscious design was a luxury, and much innovation was only possible by adapting existing resources rather than starting afresh. Moreover, I ought to acknowledge that, for many, the aesthetic qualities of an environment are more important than its functionality.

Lee (1976) describes a "well-designed study", which allocated users, at random, to two identical wards, one of which had been extensively refurbished and remodelled with partitioned bedrooms and seating areas. Direct observation, and interviews with users, showed that there was significantly more social interaction and less passive behaviour on the new ward. However, he points out a general danger in becoming too enthusiastic about this sort of intervention. There is a likelihood of confounding by the well-known "Hawthorne effect", the change in social attitudes arising from sheer participation in any kind of experiment.

Another study gives evidence that whether a therapist's room is furnished "medically" or "humanistically" affects users' evaluation of them. This interacts with the sex of the therapist, so that a man is preferred in a humanistic setting, and a woman in a medical setting (Fisher *et al.*, 1984: 310).

McDougall (in prep.), starting from the work of the architectural theorist Alexander, has begun the attempt to lay down general principles for the psychiatric milieu. His view is, that unless the spaces in a building are arranged in a sequence that corresponds to their degree of privateness, the visits made by strangers, friends, guests, users and family, will always be a little awkward. Therefore, lay out the spaces of a building so that they create a sequence. The sequence should begin with the entrance and the most public parts of the building, then lead into the slightly more private areas, and finally to the most private domains.

McDougall goes on to recommend: place the load bearing elements—the columns and the walls and floors—according to the social spaces of the building. So, never modify the social spaces to conform to the engineering structure of the building. It seems to me that this trade-off exemplifies a general tension. It is very tempting to evaluate a building in terms of its strength and durability. For there is not just the practical value of something lasting, but also we have a visceral attachment to buildings that will outlast us. Yet the social function of a building is primary, and durability may be the wrong message to send to someone in deep distress, as though their problems too may endure.

McDougall goes on to confirm that there are "counter-therapeutic" environments, which because of their complexity, distort normal patterns of living. A building that is designed to serve a simple and precise function is more likely to work effectively than a building that is expected to serve a variety of potentially conflicting purposes. There is, currently, a tendency to concentrate too great a range of services within a single building. Lee (1976), too, states that "Many writers agree that the patients in mental hospitals need clearly-defined spaces, but not so varied in function that they are a source of confusion." Since we all hope to be old one day, remember that old people have, to greater or lesser degree, restricted mobility. This physical restriction places an upper limit on the quality of the "mental map" they can form of their residence. Without a good "map", disorientation and confusion can result. Much of the confusion found in the elderly may be the product of excessively complex building design.

ASYLUM AS HOME

The word "asylum" is very negatively loaded for most people, calling to mind a host of chortling synonyms, such as "funny farm". Originally, it meant a place of refuge or sanctuary, a place to take people out of the environment that had caused their problems. "Funny farm" or not, the idea of asylum retains this crucial value and importance. The question is, whether the asylum should be as different as it is from the user's own home. I suppose there is a very weak argument that a dissimilar environment will create fewer associations to the pathogenic one. If it were possible to provide, temporarily, a *higher* standard of living, the case would be stronger. For a few people, it *was* a superior environment (and they might well become long-stay users), but not for many. The sanctuary should be a new home for the user, a home that, it is understood, is only temporary, but a home. This is a recurrent theme in the Canters' 1979 book.

McDougall resigned his post within the Health Service because of the design of the Royal United General Hospital Psychiatric Unit in Bath. The

Sunday Times (19 February 1989) took up the story, and their report touches on several fundamental issues.

> The building will be institutional, hostile, though unlike the received image of a forbidding mental hospital. Research shows that treatment centres for mental illness work best when they approximate to home. "It's my contention that the most severely mentally ill patient is still 90% normal", says McDougall. "So we must aim to mirror normal living: wake up in a bedroom upstairs, go downstairs to a living room. The World Health Organisation said that 30 years ago." ... [According to the National Association for Mental Health MIND]: "The design does not give the required sense of spaciousness and light, the openness or flexibility which contribute to personal or social relaxation, and we see some features as positively threatening". Corridors would be long and enclosed, the often-feared treatment suite for electro-convulsive therapy was tactlessly placed by the main entrance, and the single-storey layout took up too much of the site, restricting the land available for soothing gardens, MIND said. McDougall's alternative ... [was] for a two-storey complex where residential areas are separated from treatment blocks. The Health District Chairman said: "Patients will be here for a few days or a week or so. We will make the surroundings as pleasant as possible, but it's never intended to be a home"... R.D. Laing held: there's no scientific evidence to prove that a hospital-type environment is a good thing for mentally ill people, any more than it is for a woman having a baby. Nor is there scientific evidence the other way: for alternative environments.

The central question seems the "normalising" assumption that the asylum should be like home. On this question, I conclude with some words from Ittelson *et al.* (1974: 367f).

> Numerous studies over the years testify to the fact that on the order of half of all patients admitted to mental hospitals show marked improvement despite the absence of any treatment at all. For such persons the hospital may be an asylum from the stressful situation which contributed to their illness, a place in which they can gather their own strengths for another assault upon life. The environment, *qua* environment, seems to be therapy.

ANNOTATED CHAPTER BIBLIOGRAPHY

Barham P (1992). *Closing the Asylum: The Mentally Ill in Society.* Harmondsworth: Penguin.

Barham P & Hayward R (1991). *From the Mental Patient to the Person.* London: Routledge.

Chapter 2 gives thought-provoking case material about community mental users.

Barton R (1976). *Institutional Neurosis, 3rd edn.* Bristol: John Wright.

The author confesses to having been influenced by Thomas Mann's great novel *The Magic Mountain*.

Bayes K (1967). *The Therapeutic Effect of Environment on Emotionally Disturbed and Mentally Subnormal Children.* Old Woking: Unwin.
One of the earliest texts in this field.
Bean P (1980). *Compulsory Admissions to Mental Hospitals.* Chichester: John Wiley.
Mainly addressed to the 1959 Act, the approach is sociological.
Bean P & Mounser P (1993). *Discharged from Mental Hospitals.* London: Macmillan.
Beck AT (1976). *Cognitive Therapy and the Emotional Disorders.* Harmondsworth: Penguin.
Buber M (1970). *I and Thou.* (trans. W Kaufmann, orig. work 1929) Edinburgh: T & T Clark.
Bull R (1975). The psychology of clothes and fashion. *Bulletin of the British Psychological Society 28,* 459–465.
This is a good source for very accessible reviews.
Canter D & Canter S (1979a). Building for therapy. In D Canter & S Canter, *Designing for Therapeutic Environments: A Review of Research.* Chichester: John Wiley.
Includes paragraph-length summaries of all the chapters in the book.
Canter D & Canter S (1979b). Creating therapeutic environments. In D Canter & S Canter, *Designing for Therapeutic Environments: A Review of Research.* Chichester: John Wiley.
Canter D & Canter S (1979c) *Designing for Therapeutic Environments: A Review of Research.* Chichester: John Wiley.
A treasure trove for anyone wanting to pursue the content of this chapter further, and prepared to battle with polysyllables.
Ericksen MK & Sirgy MJ (1989). Achievement motivation and clothing behaviour: a self-image congruence analysis. *Journal of Social Behaviour & Personality 4,* 307–326.
Fisher JD, Bell PA & Baum AS (1984). *Environmental Psychology, 2nd edn.* New York: Holt, Rinehart & Winston.
One of the first textbooks to include the burgeoning research on therapeutic environments.
Goffman E (1968). *Asylums: Essays in the Social Situation of Mental Patients and Other Inmates.* Harmondsworth: Pelican.
The author has written other well-known books *Stigma, Relations in Public,* and *The Presentation of Self in Everyday Life.* A sociologist and a major figure.
Goldberg W & Huxley P (1992). *Common Mental Disorders. A Bio-social Model.* London: Tavistock/Routledge.
Holahan CJ (1979). Environmental psychology in psychiatric hospital settings. In D Canter & S Canter, *Designing for Therapeutic Environments: A Review of Research.* Chichester: John Wiley.
Illich I (1971). *Deschooling Society.* New York: M Boyers.
Illich I (1991). *Limits to Medicine.* Harmondsworth: Penguin.
Ittelson WH, Proshansky HM, Rivlin LG & Winkel GH (1974). *An Introduction to Environmental Psychology.* New York: Holt, Rinehart & Winston.
Jodelet D (1991). *Madness and Social Representations.* Hemel Hempstead: Harvester Wheatsheaf.
Many years of research.
Lamb R (1979). *Alternatives to Acute Hospitalisation.* San Francisco: Jossey-Bass.
Lee T (1976). *Psychology and Environment.* London: Methuen.
Gives relatively little coverage of the therapeutic environment, specifically, but what is there is good.

Leff J (1983) *Planning a Community Psychiatric Service: From Theory to Practice.* London: Royal College of Psychiatrists.
Littlewood J (1992). *Aspects of Grief.* London: Routledge.
Machiavelli N (1908). *The Prince.* London: JM Dent (orig. work 1532).
Maslow AH (1954). *Motivation and Personality.* New York: Harper & Row.
McDougall N (in prep.) The environment and mental health: the relevance of Chris Alexander's "pattern language". *Radical Health Promotion.*
McDougall is currently better known as a campaigner than a researcher, but this article promises much in terms of greater insight into the architectural needs of the psychiatric user.
NHS Health Advisory Service and DoH Social Services Inspectorate (1991) *Report on Services for Mentally Ill People and Elderly People in the Torbay Health District.*
Orem DE (1971). *Nursing: Concepts of Practice.* New York: McGraw-Hill.
Osmond H (1957). Function as the basis of psychiatric ward design. *Mental Health* (Architectural Supplement) *8,* 23–29.
Peplau HE (1988). *Interpersonal Relations in Nursing.* London: Macmillan.
Rogers CR (1951). *Client-centred Therapy.* Boston: Houghton Mifflin.
Ronco P (1972). Human factors applied to hospital patient care. *Human Factors 14,* 461–470.
Sommer R & Kroll B (1979). Mental patients and nurses rate habitability. In D Canter & S Canter, *Designing for Therapeutic Environments: A Review of Research.* Chichester: John Wiley.
Sommer is a distinguished environmental psychologist, whose article is the most useful source on this issue that I have found.
Spivack M (1967). Sensory distortions in tunnels and corridors. *Hospital & Community Psychiatry 18,* 24–30.
World Health Organisation (1953). Expert committee on mental health. *Third Report, No 73.* Geneva: World Health Organisation.
It is easy to dismiss WHO statements as utopian, witness their definition of health as "a state of complete well-being", but society needs someone to set the goals.

CHAPTER 6

Mental Retardation—Explanation and Care

Three generations of imbeciles are enough.
O. Wendell Holmes.

INTRODUCTORY

Finding a chapter title does present immediate problems of language. "Mental retardation" and "mental handicap" are familiar terms. Many would prefer them to be replaced by "learning disability" or "learning difficulty", and that is the path I am pursuing.

What seems clear is that the boundary between learning disability and mental illness is not a sharp one. Shanley (1986: 4) says the 1959 Mental Health Act resulted in "a mass exodus" from mental handicap hospitals of the most capable, whom psychiatry adopted. Among some users who remain in these institutions, something I have learned about, to my distress, is their habit of self-injurious behaviour, such as banging their heads against walls in a "perseverant" way. This seems more accurately described as disturbance, rather than learning disability. It would fit in with a well-known idea. This is that an enduring sense of helplessness (due, perhaps, to lack of normal speech) can induce depression. Perhaps it also has to do with seeking attention. The overlap with mental disturbance may explain why there has been disagreement on the incidence of "specific learning disabilities", with estimates ranging from 1 to 15% of all children.

PART A. EXPLANATION

The idea I want to combat, here, is that "mentally retarded" children are incurable. Such an idea might flow from a view that their disabilities are probably constitutional in origin. The idea of incurability was once quite current, and Rourke was an early publicist of the contrary emphasis, that they can often be trained in compensatory skills (see Vernon, 1979).

SOME TERMINOLOGICAL ANALYSIS

As will be clear by now, heredity/environment is not an either/or. Thus, one leading early scholar, J. McV Hunt, argued explicitly that intelligence has a genetic basis, yet the individual's effective ability depends on stimulation by, or interaction with, the physical and social environment.

It should be recognised, too, that there are constitutional factors present at or shortly after birth, having lasting effects, that are not inherited or genetic, but are not environmental either, in the ordinary sense.

There is an expression, "minimal brain damage", that people use in cases of dyslexia, and other learning disabilities. The adjective "minimal", here, is a technical usage. It says that there is no direct evidence, e.g., using brain imaging techniques, of physiological damage. But that since the condition cannot readily be attributed to ordinary genetic, environmental or emotional causes, a physical impairment is assumed.

Different students have brought different assumptions to the empirical evidence. Thus, a moderate correlation is consistently found between parental socio-economic status and child intelligence. I myself would automatically attribute this to the more favourable environment in which middle-class children are reared, and the deprived environment of lower-class children. Vernon (1979), on the other hand, takes the view that genetic differences between the classes are also involved to some extent. Similarly, do observed ethnic differences in intelligence show environmental or genetic effects?

There are also conceptual questions about the idea of "intelligence", some of which will become clearer after the coming section about testing it. I try to be guided, in considering it, by the belief that once it was just another neologism, an invented new term. There are major cognitive demands in understanding the literature on it, and a tempting, though mistaken, way of reacting to these is just to accept that it is a real psychological faculty. The leading academic workers on intelligence have, to a considerable extent, each defined it in their own way. For Terman (see Terman and Merrill, 1937), it is "abstract thinking ability", for Thurstone (1924) "perseverance and the capacity to inhibit instincts", while Thorndike (1966) emphasised the production of true or factual responses. Other definitions, each tending towards a different type of intelligence test, are "the ability to profit from experience and adjust to the environment", "the ability to acquire abilities", and "the capacity for knowledge". What is the reality underlying all this? I should say that the practical need to select among people for employment, as well as for education, is what drives the tests, and, to some extent, these drive the definition.

For example, I wonder how many people believe it has been discovered that average intelligence is 100? So "they" must know quite a bit about intelligence ("they" being the experts, the scientists)? I hear "But it is rather a nice round number"! Of course, there is no sort of discovery here at all.

The French government commissioned Alfred Binet (see Binet and Simon, 1908), at the beginning of the century, to find a way to identify children who could not keep up with a normal school class. He devised tests that children of a certain age could generally pass. If a child of age 8 could pass the tests for age 6 but not those for age 8, he took that to have implications for his schooling. Binet said that this child had a "mental age" of 6. He defined the child's "intelligence quotient" (IQ) as his mental age, divided by his chronological age (multiplying this fraction by a hundred to keep whole numbers). In this case, the child's IQ would be six-eighths times 100, equalling 75. Of course, the average child will have the same mental age as his chronological age. An 8-year-old should be able to do the 8-year-old tests, a 10-year-old the 10-year-old ones, and so on. The fraction to multiply by 100 will be eight-eighths or ten-tenths, so, always, one, times 100. The Binet definition of intelligence does not apply to adults, who are not so rapidly expanding their capacities. None the less, following in the child-intelligence tradition, the scores of adults, too, are mathematically "transformed" to ensure that average IQ is 100.

THE IDEA THAT THERE IS SOME "HIDDEN KEY"

It is so tempting, faced with difficult and harrowing problems, to hope that hidden away is a *key* to them. The idea, that learning disability comes from a deficiency in a unitary faculty of intelligence, is a form of "hidden-key" thinking. Stott (1971) gave a critique of the notion that the key to backwardness in school is insufficient genetic potential or the suggested global cause "perceptual deficit". He sees learning disability in terms of specific patterns or strategies of cognitive inefficiencies, induced by early environment, or linked to personality weaknesses. Among the 14 common maladaptive strategies that he lists are impulsiveness (failure to stop to think), high distractibility, and avoidance of others by withdrawal or by "playing dumb". The same children, he claims, are often quite well adapted and able in everyday situations outside the school.

Even within the world of scientific Psychology, the question whether intelligence is unitary has divided the academic community from the beginning. Charles Spearman (1927), at University College London, observed positive inter-correlations among scores on a variety of tests. He inferred these are caused by tests "loading" to varying extents on *g*, his abbreviation for "general, innate intelligence". The technical apparatus employed, which must turn off the attention of many interested laypeople, was a mathematical technique of analysing a table of intercorrelations into underlying "factors". Cyril Burt, who succeeded to Spearman's chair, claimed much of the credit for developing "factor analysis". Historians dispute his claim, which is one reason he is such a controversial figure, of which more soon. Burt's use of the technique identified further "group factors", such as "verbal intelligence" and "spatial intelligence", additional to *g*. There have been further developments. Americans influenced by the eminent worker Louis Thurstone wanted to keep the group factors, but drop the general factor. David Wechsler's (1958) widely used Adult Intelligence Scale (the WAIS) gives a verbal score and a "performance" score, and then simply combines the two to give its measure of IQ. A Scotsman, Godfrey Thomson (1951), produced another idea. He said that every test requires very many cognitive elements, and it is simply overlap of these elements that causes the positive intercorrelations. In other words, there are no factors at all. To the best of my knowledge, the "Intelligence community" has not managed to reach a consensus on this yet.

PROBLEMS IN TESTING INTELLIGENCE

Any reasonable person with a Psychology background would agree that tests yielding an "IQ" should not be followed too slavishly in deciding a need for special care. At times I myself feel that these tests are no more than parlour-games that measure idle curiosity.

However, it is possible that IQ tests have been criticised too much. It is true that children can be trained for them, to assist their chances of educational selection, but there is a limit to the success of such training. For example, vocabulary remains dependent on breadth of reading and cannot simply be expanded by training. Similarly, it is too easy to be histrionic about the extent to which testing conditions the tester's expectations, or the testees' motivation affect test scores. It is probably the large-scale group testing, that was once fashionable, that is much more liable to such weaknesses and to unreliability. The arrival of computer-based automated testing has reduced the need for that sort of testing, though negative sides to it, too, may in turn become clear. It is the theme of a provocative book by Owen (1985) *None of the Above* that the American Educational Testing Service has too much power, and that this rests upon a scientific mystique that is an illusion. A few American school systems have, indeed, now severely curtailed the use of tests to stream students.

There is now a recognition that some techniques of test-item selection restrict the range of abilities that psychologists should be concerned with. Psychologists have responded to the criticisms. Thus, they have directed much attention (too much, some have suggested) to "divergent thinking" ability (as a possible indicator of creativity). Those who think that testing is a waste of time should read the more recent theorists, such as Gardiner and R.J. Sternberg (1984). For example, Gardiner (1985) has things to say about "musical, bodily and personal" intelligence. There are also experiments investigating simultaneous cognitive performances (e.g. reading while listening to music). Whether the two performances interfere with each other suggests ways of analysing intelligence into components.

Because it is now so common, I should like to criticise what I should like to call "neuromythology" of the form that specific areas of the brain are devoted to specific types of intelligence. Any such claim is suspect, in my view, beyond the position that the cortex (surface layer) of the brain does have a special association with intelligence. In Chapter 1 I mentioned "phrenology" as a historical movement, and the idea of identifying areas of the brain that govern specific faculties has an enduring appeal. It receives a strong boost after episodes of armed conflict. Then, "clean" penetrating missile wounds of the brain present to doctors. With the usual pathologies of stroke, tumour or epilepsy, it is difficult or impossible to specify the location of damage exactly. World War II left many people, some alive today, with localised brain damage. Having read in this field, I feel convinced that "integrative" functions such as intelligence, or even memory, depend on the activity of the brain as a whole, its "mass action". Once, we believed, at least, that verbal intelligence is associated with the left hemisphere of the brain, for example, Reitan (1955, 1966) did some work in this vein. Currently, each year seems to bring forth further evidence of the importance of the right hemisphere, too, for language. Quite early, Eisenson (1962) said that there *was* a speech impairment of right-brain-damaged patients, but not impairment that would become apparent to the clinician in ordinary conversation. The impairment would manifest itself, for example, in excessive circumlocution. One landmark in the developing reappraisal, has been a *Psychological Bulletin* review article by Searleman in 1977.

I have sometimes noticed signs of deep feeling about the issue of intelligence. It was *The Intelligence Man*, who has provoked the greatest controversy to beset British Psychology so far. With this sobriquet a BBC documentary, broadcast in 1980, described Sir Cyril Burt.

Cyril Burt was knighted for his services to Psychology. The roll-call of distinguished British psychologists goes back to the beginning of the century, with W.H. Rivers, C. Spearman and C.S. Myers, or even earlier, to Sir F. Galton (of whom more later). Burt, who rose to prominence during the inter-war years, won the first post for a practical psychologist. He worked with the London County Council to advise on children's education. He managed to combine this with learned publication. Perhaps, in view of this workload, it is not so surprising that his marriage became unhappy. He turned ever more to academic life, and achieved great influence and eminence.

Some years after his death, there came a major upset, to psychologists and others. An American psychologist at Princeton University, Leon Kamin (1974), accused him of having written deliberate lies, of having made fraudulent claims in his articles. Many found the allegations preposterous; but they would not go away. A leading medical journalist, Oliver Gillie, took up the charges, and the Clarkes, students of Burt and Professors of Psychology at the University of Hull, endorsed them in Gillie's newspaper articles. Burt's first biographer, L. Hearnshaw (1979), reluctantly concluded, on sifting the evidence, that the charges had substance, and the British Psychological Society stated that, on balance, they were justified. The BBC film was the first and last most people have heard of it, but within the academic world the controversy will not lie down. R.B. Joynson (1989) and Ronald Fletcher (1991) have written recent books to vindicate Burt, and the British Psychological Society has withdrawn its previous statement. My own feeling on this subject, is that Burt did make errors in print, but more because of the cognitive effects of age, than through any intention to deceive. It is a fact that there is much scope for error in this sort of work.

What I want to highlight about the Burt controversy, is that it occurred in the area of Intelligence. This is an area that arouses particular public attention, reflected in media coverage. Normally, academics plough their own furrow, without the pressures of this sort of scrutiny.

PERI-NATAL EVENTS

There seems little doubt that the cause of learning disability can sometimes be found, if not in genes, in very early, i.e., peri-natal events. Pregnancy complications, premature births and mothers who are much above the average childbearing age, are acknowledged risk factors for learning disability, as well as for epilepsy, cerebral palsy, reading disability and certain behaviour disorders (Vernon, 1979). Vernon also cites Gottfried and Pasamanick for the view that learning disabled children have often been affected by peri-natal anoxia (oxygen deficiency).

Stott (1971) consulted the medical records, and interviewed the mothers, of 105 "subnormal or mentally defective" children. He found among 49% of them, maternal illness or emotionally stressful conditions, such as matrimonial problems, housing difficulties or illness in the family, during pregnancy. He similarly questioned the mothers of a control group of mentally normal children (including some siblings of the subnormals); only some 15 to 20% of these women had comparable relevant pregnancy difficulties. The difference in incidence could not be accounted for simply by differences in poverty between the two groups; the difference in conditions of stress was crucial.

When it comes to peri-natal nutrition, however, which seems an obvious possible risk factor, the evidence is unclear. Stein *et al.* (1972) tabulated the test scores, at age 19, of some 20 000 Dutch army recruits, whose mothers had been subjected to severe undernourishment during the peri-natal months, at the time of the German occupation in 1944–45. Compared with a 100 000 recruits, whose mothers had not so suffered, there was little suggestion of learning disability. A WHO report (1974) says that "there is practically no evidence of a relationship between . . . mild and moderate forms of malnutrition and mental retardation. What seems more probable is that there is an interaction between malnutrition and other environmental factors" (see Vernon, 1979). The situation, however, may be very different with regard to severe malnutrition in underdeveloped countries.

DEPRIVED UPBRINGING

The textbooks (e.g. Shanley, 1986: 16) assert that mild learning disability is most prevalent within social classes four and five. With this can go a typical explanation. This is that there is a continuing cycle of poverty, leading to peri-natal abnormalities, leading to children who fail at school, so are only fit for low-grade employment and live in poverty, and so the cycle repeats itself. Animal studies show that early deprivation (typically, in this experimental work, of an extreme form, in human terms) leads to learning disability. This is not comparable to any but occasional examples, of radical social and physical deprivation during human childhood, certainly not to mere low socio-economic status. Old studies, by Gordon (1923), of canal-boat and gypsy children, also suggested their lack of schooling reduced their IQ. Some scattered cross-cultural evidence, too, suggests that complete absence of schooling seriously retards mental growth. One study, of Indian children in South Africa, showed that lack of schooling during ages 7–9 produced a learning disability equivalent to 5 IQ points a year.

LONGITUDINAL VARIABILITY

Several writers have claimed that the scholastic retardation of disadvantaged children is cumulative. Because they are backward when they first come to school, they do badly from the start; progressively, they become more frustrated and disheartened and, therefore, fall more and more behind in later grades. If confirmed, this hypothesis would imply that the longer the exposure to a deprived or non-stimulating environment, the more IQ and achievement might be expected to fall below the average. On genetic theory, on the other hand, we might expect low-IQ children to show a deficit in school work as soon it depends on conceptual learning.

This would then persist with little alteration (other than individual fluctuations) till the end of their school careers. Vernon (1979) has no doubt that the deficit in ability and achievement does become more pronounced with age.

GENETICS OF INTELLIGENCE

The genetics of intelligence is an area beset by controversy, and, perhaps consequently, lack of clarity. In our own time, Leon Kamin, a distinguished psychologist who began his career in the study of animals, has waged unrelenting war against the idea of *any* genetic influence upon IQ. A generation before him, Quinn McNemar, author of a standard text on psychological statistics, lent himself to an equally wholesale debunking of any *environmental* influence on IQ (see Snyderman and Rothman, 1988: 98).

Three points congruent with the genetic view seem right to me. First, certain types of pathological mental defect are, indeed, determined by chromosomal anomalies, for example Down's syndrome and Turner's syndrome. The abnormal number of people with IQs below 50 does point to some pathological deficiencies. Unlike those with IQ 50–70, there is little or no association in this "pathological tail" (of the distribution) with low socio-economic status or with below-average relatives. Second, "inbreeding depression" of intelligence frequently occurs with incestuous matings. This would be expected on genetic grounds, and appears intractable to environmental explanation. Researchers have found a (smaller amount of) depression in intelligence in first-cousin marriages, but there are difficulties in avoiding biased sampling. Third, we know from animal breeding experiments that skills, as well as physical features, can indeed be selectively bred.

However, there are some common fallacies about the genetics of intelligence. The distribution of IQ in the general population is the "normal" bell-shaped curve. But this does not prove polygenic inheritance. The shape of the distribution is, to some extent, under the control of the investigator, anyway, but any score determined by multiple small randomly distributed factors will follow the normal curve.

There is another fallacy. One phenomenon found in family studies of IQ is "regression to the mean": children are usually closer to the mean of 100 than their parents are. This, too, says nothing about a genetic mechanism, but is simply an effect of imperfect correlations. Children are likely to be nearer the mean of the IQ distribution than their parents because that is where the bulk of the normal distribution lies. It is a telling point, in the polemics over the heritability of IQ, that Kamin knows this, but Hans

Eysenck appears not to. I base this opinion on their joint book on the issue, *Intelligence: Battle for the Mind*.

The trend in modern Psychology to study very early experience has revealed the limitations of geneticism. As Gillie (1976: 177) says:

> A baby learns to understand language and to speak from its parents, and this understanding forms the basis of learning in school. If the parents have a poor understanding of language or do not talk to the child, then the child cannot learn, however bright it is. A mass of mathematical theorising by hereditarians such as Jensen, Herrnstein and Eysenck has obscured this basic fact, promoting a dogma that intelligence is 80 percent inherited.

Even among these psychologists of the behaviour genetics school, the percentage of intelligence attributed to genes does seem to be coming down to around 60%. Vernon (1979: 199f) gives the view:

> Several plausible components of [IQ] variance have been distinguished, including assortative mating between parents, dominance (or gene interaction), genetic–environmental [GE] covariance and interaction, and differences in effects between families and within families. The available data on different kinship groups are quite inadequate to sort out so many variables. GE-covariance would appear particularly important, but it has been ignored in some analyses . . . There appears to be some convergence from the different researches on roughly 60% genetic, 30% environmental and 10% GE-covariance. But in view of all the uncertainties, it is unlikely that general agreement will be reached on any precise figures.

Vroon (1980: 110) says, "Depending on their preferences regarding estimated parameters, on the ground of the data concerning IQ concordances for identical and for fraternal twins, [behaviour geneticists] have come to a figure between 15% and 72%". In the next chapter I say much more about the problems with the kinship studies. The disagreement about the figure is also caused by the fact that there are different sorts of genetic model. Dr Robert Goodman of the Institute of Psychiatry in London says that as few as five genes could each account for 10% of the variance in intelligence. On the other hand, he acknowledges there could be 1000 genes each accounting for 0.1% of the variance (polygenic inheritance).

The figure of 60% is an estimate of "heritability". There are some difficult points that have to be grasped about heritability. It is important to realise, that it is defined as the proportion of *variation* in a trait attributable to genetic variation. It is a population-specific statistic. In a different population, or the same population at a different time, a different heritability estimate may be obtained for the same trait, as genetic and environmental variation themselves change. A very high heritability for IQ does not mean that IQ cannot be altered through environmental change. Even height, which is highly heritable, has increased two standard deviations (12 cm) in the last century. What high heritability does mean is that, under the present circumstances, an environmental change that is large, relative to present levels of environmental variation, is needed to produce substantial changes in IQ.

We are looking back on a century of research of this sort. Some believe that the Human Genome Project, the worldwide collaborative attempt to discover the complete sequence of bases within human DNA, will reveal all. Hope beats eternal.

HANDEDNESS AND INTELLIGENCE

I have myself made one empirical research effort to show a link between biology and intelligence. I conclude this first section on the *explanation* of learning disability by reviewing it.

Most people show a bias towards one hand, more frequently the right, but for about one in ten the left. For about a century, people have thought left-handedness has neurological significance (see most of the articles in Coren, 1990). They often describe it as a "soft" neurological indicator, i.e., a sign, but not a reliable one, that someone has a neurological problem of some sort.

The evidence is better that left-handers are at least neurologically *different* from the majority. For some time after Broca (1861) (following Dax—see Benton, 1964/5) proposed that speech is handled principally by the left hemisphere of the brain, it was supposed that left-handers would show, conversely, right cerebral "dominance". For we know, that many incoming nerve pathways cross over from the lateral side (left or right) of the body to the opposite side of the brain. "Left-handedness" was often taken, then, to mean writing (a linguistic activity, if very different from speech) with the left hand. This view, that left-handers show "crossed" dominance, was overturned. There is a test involving "anaesthetisation" of one hemisphere exclusively, by injecting sodium amytal into one carotid artery. This revealed, with a large number of left-handed epilepsy sufferers, that a majority were still left-hemisphere-dominant, though it was more like a 70% proportion than the 98% of right-handers (Milner *et al.*, 1966; a good review is Milner, 1972). They also differed in that, for many of them, *both* hemispheres seemed to play a part in speech. This would account for evidence that left-handers are more likely to have difficulties with speech ("dysphasia") following brain injury (side of injury not exclusive or not specified), but are also more likely to recover from it quickly (Hécaen & Piercy, 1956).

It was in 1969, that Jerre Levy suggested that the tendency towards bilateral organisation of language in left-handers would interfere with and depress their performance of non-verbal tasks, though it might give them a slight advantage on verbal tasks. She gave the Wechsler Adult Intelligence Scale to 10 left-handers and 15 right-handers (all graduate students), and found that the difference score (Verbal IQ minus Performance IQ) for this test was greater (statistically significantly) for the left-handers than for the right-handers. This difference score was due mostly to left-handers' lower non-verbal IQ. Her reasoning was, that non-verbal IQ depends more on the right than on the left hemisphere (Blakemore *et al.*, 1972), and in left-handers this hemisphere is not freed for this form of intelligence, exclusively, so the score goes down. There are two issues arising: the reliability of Levy's result, and the validity of her reasoning.

> The teasingly suggestive nature of this relationship [between left-handedness and deficit] is rather striking in its persistence through generations of research workers. There is usually just enough of a relationship to suggest a possible link and never enough of one to establish firmly a solid correlation

said Hardyck & Petrinovich (1977: 393f). I have conducted research of my own (Williams, 1987c) on this topic and endorse this statement, both from my own data and from my review of more recent studies. At the end of this chapter, there is a separate bibliography on Handedness and Deficit, drawn from this review, with numerous further references as well. Some references are books providing their own reviews, and some are empirical studies with much larger samples than Levy's 25.

The second issue is the validity of Levy's reasoning. Just because language and non-verbal functions are competing, in some sense, for the use of the right hemisphere in left-handers, this does not mean that the non-verbal intelligence will be the one to be impaired. It could just as well be language that shows a deficit. Indeed, during the inter-war period, researchers concentrated on the possibility of a language difficulty, such as what became known as "dyslexia", in the left-hander.

I regard this handedness research area as an illustration of the mistaken bias towards physical, neurological, genetic explanations of deficit. It is the mere fact, of being left-handed in "a right-handers' world", (see every issue of *The LeftHander*; Figure 6.1). Being part of a minority group, with the age-old pejorative associations attached to the "sinister" hand, is enough to explain the small inferiorities shown. As an illustration of how small they are; in my own 1987 study, handedness accounted for less than 3% of individual variation on the measure used. Writing from left to right is more difficult, mechanically, for left-handed children. Unless they hook their hands round, perhaps awkwardly, so that they can *pull* a pen across paper, they obscure what they have just written. Implements, machines and forms tend to be designed for right-handers. Very often, even today,

Figure 6.1 It's a right-handers' world. Reproduced by permission of Eliza Kentridge

children who prefer to write with their left hand may have to weather their teacher's disapproval. Burt (1950) advised compelling such children to use their right hand. What sort of emotional disturbances may this not cause?

At the same time, I feel that the very minority status of left-handedness lends it a genuine interest. Sometimes children suffer a major sporting accident that both switches their preferred hand and also disrupts their schooling. Logically, such children will form a bigger proportion of left-handers than of right-handers, just because the former is a smaller group. In other words, there may be a substantial number of "pathological" left-handers. Essentially the same argument has been adapted to "neurological" switching. Suppose "minimal brain damage", perhaps occurring as early as peri-natally, can lead to someone switching their preference to the hand they would not otherwise have used. Then such people will form a bigger proportion of left-handers (Satz, 1973). In this way, some sort of deficit for left-handers *as a group* (most of them will be entirely typical) can be understood.

PART B. CARE

SCENE-SETTING

Even after World War II, 21 States of the Union practised sterilisation of learning-disabled people, the basis for selection being their IQ scores (Shanley, 1986: 3). For ever, the authorities were barring them from the joy of parenthood. It takes little imagination to realise that this is just the tip of an iceberg of man's inhumanity to man, with parenthood being barred in other ways, and with other forms of discrimination. The IQ test is constructed so that 2% of the population will score below 70 points, and that was the standard cut-off for official registration in Britain. But in such circumstances, it is hardly surprising that a study of the official register for Sheffield (England) in 1976 found only a quarter of that percentage actually recorded (Shanley, 1986: 9). It is patently obvious that many will not seek help for themselves, or for their relatives, when stigma, and its consequences, have been so grave in the recent past.

Consider the succinct expression of stigma and discrimination by the illustrious jurist Oliver Wendell Holmes, in the Supreme Court of the USA, in 1927: "Three generations of imbeciles are enough" (see Gould, 1985). Perhaps it is unfair to represent this eminent American by that one phrase. I have chosen it because it does, of course, have a double meaning. Society has neglected this scourge too long; more public attention and resources should be given to learning disabilities. It is in the spirit that my attention is something, even where expertise is lacking, in that I am writing this chapter.

Gould (1985) has discussed this 1927 case of compulsory sterilisation tellingly and with great power. It was the case of Carrie Buck. In the legal proceedings, her Stanford–Binet mental age was said to be 9 years, that of her mother, Emma, 8 years. For the third generation, the evidence of imbecility was less scientific. In court, a social worker said "There is a look about [Carrie's existing daughter Vivian] that is not quite normal, but just what it is, I can't tell". In 1980, the director of the hospital, where Carrie had, on Holmes's judgment, finally been sterilised, researched the records of his institution. He found more than 4000 sterilisations had been done, the last as late as 1972. One of them was Carrie's sister, Doris. She had been told the operation was for appendicitis, and only found out in old age why she had never been able to conceive, something she had wanted more than anything else. The critical and shocking fact was that when the case was publicised again, all these years later, it was patently clear that Carrie's intelligence was normal. It turned out that the reason for her institutionalisation was that she had been raped by a relative of her foster parents, then blamed for the resulting pregnancy, and committed to "hide her shame" (and the rapist's identity). All journalists and professionals interviewing her were convinced that she was of normal intelligence. Yet, despite having her case taken to the Supreme Court, she had been sterilised, to avert the possibility of a "further generation of imbeciles".

A positive act of sterilisation is, as I said, only the tip of an iceberg. Records on the easily reversible forms of contraception are not readily available. Even if they were, institutional living places inherent obstacles in the way of the joy of sexual intercourse. Until opportunities are provided that are obvious enough to these people for them to take them, it can be said that "eugenic" policies are being pursued (see Chapter 7 for a discussion of eugenics).

In this context, is the following statement, of Vernon, in 1979, surprising?

> Psychologists regard their work as being as confidential as a doctor's files. A psychologist could hardly examine or treat a retarded or emotionally maladjusted child without recording comments about the home that the parents might resent. It has been an almost universal policy that parents should not be told their child's IQ, because they are liable to misinterpret it.

In the UK, it is now official policy, that all doctors', psychologists' and nurses' records should be open to the user. No-one who has actually seen such records could believe the genuine implementation of this policy would lead to anything other than a great surge of anger and resentment. This whole area is shrouded in an oppressive cloak of silence. It is easy enough to understand why that is so: IQ carries emotional baggage. Most of us are sensitive about our own IQ. Sensitivity is the feeding-ground for humour. The BBC is supposed, since Lord Reith, to set cultural standards for the nation. Yet it recently ran a radio billboard advertisement that read, "If your IQ is lower than our frequency then we're not 4 you. Radio 4—93.5 FM". What heartens me is that this misplaced humour was taken up by the charity Mencap, who had the advertisement dropped, and perhaps have changed attitudes at the BBC a little.

Care for individuals is made necessary by the social failure to prevent, if you accept the argument of Part A of this chapter. "People find it comforting to believe in genetic differences, since it saves them doing anything about depressed minorities or social reform", says Vernon (1979), who does not seem to me a rabid revolutionary. Attention to this subject, more than most, seems to castigate the apathy or even self-interest of those who wish to keep society as it is.

PROGNOSIS

As I have said, heritability is a population statistic, not a fixed property of intelligence. That is, it depends on the group being studied. It goes down when the range of environments experienced by the group is large, and increases when environments are relatively homogeneous. High heritability does not imply that teaching and learning are unimportant, only that the range of variations in an ability produced by *currently available* educational and child-rearing environments is limited.

Another reason for downgrading the potential of education, is the idea of a "critical period" during early infancy. Then, it is said, the absence of particular experiences, especially of the formation of a close attachment to the mother, can lead to irreparable psychological harm. The idea bears a close relation to the phenomenon of "imprinting" described by Konrad Lorenz (1981). He found that goslings will treat as a mother (e.g., by following) *any* object, such as a long yellow glove on the arm of an experimenter. D.O. Hebb, an eminent American post-war psychologist, applied the idea enthusiastically to humans, but with time the early period came to be described as "sensitive" rather than "critical". Then the Clarkes (Clarkes and Berg, 1985) (whom I have mentioned before in the Burt affair), in particular, argued that undue credence had been given to the primacy of early experiences over later learning. This is analogous to one divide between Freudian and Jungian analysis.

The other major point I should like to make here is that parental education and intellectual stimulation in the home are recognised as generally more predictive of child ability than are material circumstances.

IMPROVED UPBRINGING

Occasionally, children are discovered who have been abandoned in the wild and helped to survive by animals (for which reason, they are called "feral" children). There are cases, too, of children discovered after spending their lives in a grossly abnormal environment. One was "Jeanie", an American, who had lived her life until discovery in a cellar, with no human contact except to receive food. Though such children are typically highly learning disabled, it seems they can usually be brought back to normal by removing them to a suitable environment, even in middle childhood. There are many more children from dramatically poor environments who have not received the same media attention. A classic long-term follow-up study by Skeels (1966) showed that palliating the effects of institutional life by fostering such children can produce IQ gains of 30 points and over.

A strong environmentalist trend dominated US society, and so its Psychology and Sociology, in the 1950s and 1960s. There was a general belief that compensatory early schooling could overcome the effects of deprived environments on children's educational achievement. In particular, the well-known "hero project", called Head Start, was set up. The general perception of the Head Start programmes as a failure touched off the reaction given a written canon by Jensen (1969).

There was a major *experiment* by Heber (1962) on remediation, beginning at the age of 2, of intellectually and educationally disadvantaged children. This resulted in consistent IQ differences between experimental and control groups of 20–30 points. The experiment involved training mothers to interact more effectively with their preschool children, as well as intensive stimulation of the children's cognitive and linguistic skills. Though the programme ceased on entry to elementary school, the experimental group still scored over 20 points higher than the controls on long-term follow-up.

Many additional intervention experiments with young children, usually involving mother–child interactions, have shown substantial, and apparently lasting, gains. But studies of ordinary nursery or pre-school classes for children below 5 suggest them to be rather ineffective, perhaps because they are often no more than a way of caring, for part of the day, for children with career mothers.

INSTITUTIONS

In practice, many people with learning disabilities live in institutions, often large hospitals with a wide catchment area, a social answer sometimes described as "warehousing" (Figs. 6.2 and 6.3). Institutional life probably exacerbates their disabilities. Vernon gives the opinion that children separated from their parents, and reared in typical institutional surroundings, tend to be somewhat retarded anyway. I believe that the numbers on a ward are a main hindrance to being able to give care. It seems that many dormitories are now smaller, more private and more personal (Shanley, 1986: 79), with attempts at more homely furnishings.

The ward is not only a home, but also the organisational centre for treatment programmes. Goffman's (1968) ideas on total institutions are again very relevant. One facet he emphasises, is "role dispossession" of the user (having no opportunity to present different selves to different groups). The user receives uniform treatment, which is depersonalising. There is a split between staff and residents (see Shanley, 1986: 12). This sort of work, for which it is so difficult to organise effective acknowledgement, readily burns out staff, who will then tend to stay in their own rooms.

Figure 6.2 The predecessor to Turner Village Learning Disabilities Hospital

Figure 6.3 Turner Village Hospital (central ring in aerial view). Reproduced by permission of Medical Photography, Essex County Hospital, Colchester

REMEDIATION

Rather than *monuments* to concern (or indifference), "community" options for a care environment for the learning disabled are becoming more common. These include children's homes with reserved places, fostering, hostels, group homes, sheltered housing, lodgings and nursing homes.

Many hold that the development of oral verbal skills is more important than reading, for such children. The contention, in Piaget's terminology, is that learning-disabled adolescents are unlikely to reach the formal operational stage of development. Shanley (1986) says that they learn by manipulating and exploring, rather than by being taught how to do things, and he stresses the importance of play as a source of "incidental learning".

The metaphor of "warehousing" should be replaced by one of "horticulture". Vernon believes the apparent ineffectiveness of programmes such as Head Start may have been due, partly, to poorly chosen objectives and methods, and to the lack of adequate criteria for evaluation. He is an advocate of the "mastery learning" approach, put forward by Bloom (1975). The main ingredient of this is a clear formulation of the basic objectives of the course, divided up into a series of subtopics. It seems likely that the provision of frequent feedback (on whether the user has mastered a subtopic) will be very beneficial both for user and teacher.

THE PHYSICAL ENVIRONMENT FOR CARE

Perhaps it bears repeating, that the social environment of care is far more important than the physical environment, but for reasons already discussed I am going to concentrate on the latter.

M. Jones (1983) has written a book about Beech Tree House, a home for learning-disabled children with behaviour problems. Something he writes (p51) is unsurprising but worth reproducing.

> Most of the rooms have little that is special about them ... [but] we endeavoured to keep all rooms as bright and as pleasant as possible by using attractive wallpaper, colourful curtains, gay bedding and furnishing, and by displaying children's personal photographs, work and pictures on the walls.

The house does have some special features: the Multipurpose Room, and a Time Out cubicle in the classroom. Jones points out, that some children arriving at the house need a low-distraction environment for initial learning. One of the most potent distractions for children in group teaching situations is the acting out behaviour of other children. This leads him to assert the value of partitions. He also takes the unpopular view that a locked door policy is justifiable, because it releases staff from purely supervisory duties. In mental health, it is common to proclaim "the drug revolution has ended the locked ward". Do the horrors of the old back ward really mean there is never a place for locks? Lack of staff time is the problem that greatly exercises Jones, as a practitioner. Elsewhere, too, he says (p26) "Wherever possible we designed Beech Tree House in a way which minimised the time spent in purely supervisory duties". At the same time, there is always a danger, in this sort of work, that the convenience of the staff may be in tension with the needs of the user. Clients may spend all their time segregated away in their own lounges, while nurses stay in a glass-enclosed station. The various services, such as classrooms, therapy areas and hair and beauty shops, may be located inside the residential building, rather than to give users a change of setting and a sense of home. Frequently, in documentation, one encounters terms like "abuse-resistant", or "damage-proof", or "heavy-duty" environments. "Purpose-built itself, though it can just mean "not an old asylum", also often implies abnormal and non-home in some way. It has been suggested that the better course is to adapt existing community housing. Even Building Codes are often unhelpful for care purposes.

Wolfensberger (1972) points out that some buildings used for this sort of care look like prisons, even having bars on the windows. Sometimes they have been taken over from other users with behaviour problems. The effect is, naturally, to deter other people in the community from interacting with the residents. If outsiders do venture in, they may be greeted by a non-enclosed toilet. The environment gives users no opportunity to enjoy a wide range and large number of "normalising" experiences. Wolfensberger believes that a neighbourhood or community can only integrate so many deviant persons at any one time. So the size of a facility should be such as to congregate no more deviant users than can readily be absorbed in, and by, the surrounding area, services, resources, social life, etc. Smallness of size, in turn, dictates that residential services should be specialised for specific types of problems or groups (p36).

Husbands whose wives are returning to work after having children may discover anew how they are taking for granted skills around the kitchen. Much time and energy may be invested in teaching users kitchen skills. Yet the Gunzbergs (1979) point out that it is not reaping its full harvest unless the physical environment provides an accessible kitchen, where users can exercise and develop the learning. They also say that even the width of corridors the user regularly encounters is a two-edged factor. It is natural to assume that more space is desirable, but a narrow corridor may actually help to teach the user some social niceties (how to handle passing by someone else, or between two people having a conversation).

In passing, let me say that the phrase "social skills" is entering the language, and indeed it is the focus of a major academic effort. I should like to mention in particular the work of Roger Ellis, and of Owen Hargie. Yet, like many modern movements, it has its precursor in folk wisdom. "Manners maketh man" is the motto of an English public school, and the value of civility is difficult to overestimate. As a psychologist myself, I know and regret the pitfall of making personal remarks, into which other people's expectations of the profession can lead me. It may sound trite, but I believe one social skill with strangers is to restrict personal feedback to peripheral, rather than core aspects, of someone's self-concept (don't get *too* personal). Resist the temptation to issue forth profundities, which are often hollow as well as hurtful, or to project mystique.

Another of the Gunzbergs' themes is "Do not make things too easy for the user". Thus, it seems considerate, to arrange the chairs in a television lounge appropriately for the number of residents. On the other hand, making a user fetch a chair from somewhere else, and then return it to where he found it, will teach him domestic habits that will be more valuable to him in the future. The Gunzbergs (1979) endorse Wolfensberger's theme of "normalisation", and make some other detailed points in the same vein. They summarise as follows (p171),

> The problem in many ways is simply that of realising that the physical environment can be a therapeutic tool if the treatment philosophy accepts that treatment in "learning to live skills" requires the right type of environment.

They emphasise that this approach is not a magic wand, to wave over sometimes daunting problems, but merely a step in the right direction.

ANNOTATED CHAPTER BIBLIOGRAPHY

Benton AL (1964/5). Contributions to aphasia before Broca. *Cortex 1,* 314–327.

Binet A & Simon T (1908). The development of intelligence in the child. *Année Psychologique 14,* 1–90.

Blakemore C, Iversen SD & Zangwill OL (1972). Brain functions (asymmetry of cerebral hemisphere function). *Annual Review of Psychology 23,* 433–456.

Bloom BS (1975). Mastery learning and its implications for curriculum development. pp 334–350. In M Golby, J Greenward & R West (Eds) *Curriculum Development.* London: Croom Helm.

Broca P (1861). Remarques sur le siège de la faculté du langage articulé, suivies d'une observation d'aphème. *Bulletin de la Société Anatomique de Paris 6,* 398–407.

Burt C (1950). *The Backward Child, 3rd edn.* London: University of London Press.
It is difficult to believe that the author of this is any kind of charlatan.

Canter D & Canter S (1979). *Designing for Therapeutic Environments: A Review of Research.* Chichester: John Wiley.
Several articles in this landmark collection are relevant to learning disability.

Clarke AM, Clarke ADB & Berg JM (1985) (Eds) *Mental Deficiency: The Changing Outlook, 4th edn.* London: Macmillan.

Coren S (1990). *Left-Handedness: Behavioral Implications and Anomalies.* Amsterdam: North Holland.

Like many edited books a bit of a lucky dip. Many of the contributors have established reputations in this field. I recommend the chapters by Peters and by Harris.

Eisenson J (1962). Language and intellectual findings associated with right cerebral damage. *Language & Speech 5,* 49–53.
A strangely neglected article.

Ellis R & Whittington D (1986). *A Guide to Social Skills Training.* London: Croom Helm.

Eysenck HJ & Kamin LJ (1981). *Intelligence: The Battle for the Mind.* London: Macmillan.
Cast in the form of a debate, with each author both proposing and seconding in a separate section his view.

Fletcher R (1991). *Science, Ideology and the Media: The Cyril Burt Scandal.* New Brunswick, NJ: Transaction.

Gardiner H (1985). *Frames of Mind.* London: Paladin.

Gillie O (1976). *Who Do You Think You Are?* London: Hart-Davis.

Gordon, H (1923). Mental and scholastic tests among retarded children. *Board of Education Pamphlet No 44.* London: HMSO.
An early classic.

Gould SJ (1985). *The Flamingo's Smile: Reflections in Natural History.* London: WW Norton.

Gunzberg HC & Gunzberg AL (1979). "Normal" environment with a plus for the mentally retarded. In D Canter & S Canter, *Designing for Therapeutic Environments: A Review of Research.* Chichester: John Wiley.
Gives a collection of further readings, though without authors named.

Hargie O, Saunders C & Dickson D (1987). *Social Skills in Interpersonal Communication.* London: Croom Helm.

Hearnshaw LS (1979). *Cyril Burt, Psychologist.* London: Hodder & Stoughton.

Hebb DO (1987). *A Textbook of Psychology, 4th edn.* by DC Donderi. Hove: Lawrence Erlbaum.

Heber R (1962). The concept of mental retardation: definition and classification. *Proceedings of the London Conference on Scientific Studies of Mental Deficiency 1,* 236–242.

Hécaen H & Piercy M (1956). Paroxysmal dysphasia and the problem of cerebral dominance. *Journal of Neurology, Neurosurgery & Psychiatry 19,* 194–201.
Two very distinguished names in clinical neuropsychology.

Hunt J McV (1961). *Intelligence and Experience.* New York: Ronald Press.

Jensen AR (1969). How much can we boost IQ and scholastic achievement? *Harvard Educational Review 39,* 1–123.
Why did Jensen publish this in his own University's journal? It is the article for which he was reviled as a racist and ultra-conservative. For years afterwards he was unable to give a talk in any university without smoke-bombs and general disruption.

Jones MC (1983). *Behaviour Problems in Handicapped Children: The Beech Tree House Approach.* London: Souvenir.
The author is diffident about his own writing abilities but has a good story to tell.

Joynson RB (1989). *The Burt Affair.* London: Routledge.

Kamin LJ (1974). *The Science and Politics of IQ.* Potomac, MD: Erlbaum.
Once, I was horrified by this title, regarding politics as a mucky business that should be kept separate from objective questions. Now, I understand more about the feelings involved. None the less, Professor Kamin applies research standards

that are only feasible in the controlled conditions of the animal laboratory to a question that does have to be researched without fear of personal hostility.
Kavanagh JF (1988). *Understanding Mental Retardation: Research Accomplishments and New Frontiers.* Baltimore, MD: Paul Brookes.
Lorenz K (1981). *The Foundation of Ethology.* New York: Springer-Verlag.
McNemar Q (1955). *Psychological Statistics, 2nd edn.* New York: Wiley.
Medawar P & Medawar J (1977). *The Life Sciences.* New York: Harper & Row.
Lord Medawar was a Nobel prizeman, making fundamental discoveries on the reasons for immunological rejection of tissue grafts. Many people owe their lives, their appearance, and probably their sanity to him.
Owen D (1985). *None of the Above: Behind the Myth of Scholastic Aptitude.* Boston: Houghton Mifflin.
An angry book, that must leave the reader with anxieties about the validity of testing for educational selection in America.
Reitan RM (1955). Certain differential effects of left and right cerebral lesions in human adults. *Journal of Comparative and Physiological Psychology 48,* 474–477.
Reitan RM (1966). Problems and prospects in studying the psychological correlates of brain lesions. *Cortex 2,* 127–154.
Searleman A (1977). A review of right hemisphere linguistic capabilities. *Psychological Bulletin 84(3),* 503–528.
It took some cussedness to write a review like this at the time.
Shanley, E (1986). *Mental Handicap: A Handbook of Care.* London: Churchill Livingstone.
A good standard textbook.
Skeels, HM (1966). Adult status of children with contrasting early life experiences: a follow-up study. *Monographs of the Society for Research in Child Development 31, No. 105.*
A pivotal study.
Snyderman M & Rothman S (1988). *The IQ Controversy: The Media and Public Policy.* Oxford: Transaction.
Spearman CE (1927). *The Abilities of Man.* London: Macmillan.
Stein Z, Susser M, Saenger G, & Marolla F (1972). Nutrition and mental performance. *Science 178,* 708–713.
Sternberg RJ (1984). Towards a triarchic theory of human intelligence. *The Behavioral and Brain Sciences 7,* 269–287.
Stott DH (1971). Behavioural aspects of learning disabilities: assessment and remediation. *Experimental Publications System 11,* 400–436.
Terman LM & Merrill MA (1937). *Measuring Intelligence.* London: Harrap.
Thomson GH (1951). *The Factorial Analysis of Human Ability, 5th edn.* London: London University Press.
Thorndike RL (1966). Intellectual status and intellectual growth. *Journal of Educational Psychology 57,* 121–127.
Thurstone LL (1924). *The Nature of Intelligence.* London: Kegan Paul, Trench, Trusner.
Vernon PE (1979). *Intelligence: Heredity and Environment.* San Francisco: WH Freeman.
A comprehensive monograph written by an academic psychologist who spent his career researching intelligence. This is the source you should go to for more depth on the issues of Part A of this chapter, which owes a great deal to this book.
Vroon PA (1980). *Intelligence: on Myths and Measurement.* Oxford: North Holland.

Reasonably slim, and full of interest.
Wechsler D (1958). *The Measurement and Appraisal of Adult Intelligence.* London: Bailliere, Tindall & Cox.
Wolfensberger W (1972). *Normalization: The Principle of Normalization in Human Services.* Toronto: National Institute of Mental Retardation.
This was the origin of the normalisation movement. The prose can be pretty heavy but some interesting insights can be readily decoded.
Zigler E & Hodapp RM (1986). *Understanding Mental Retardation.* Cambridge: Cambridge University Press.

HANDEDNESS AND DEFICIT BIBLIOGRAPHY

Annett M & Manning M (1989). The disadvantages of dextrality for intelligence. *British Journal of Psychology 80,* 213–226.
Annett M & Manning M (1990). Arithmetic and laterality. *Neuropsychologia 28,* 61–69.
Annett M & Turner A (1974). Laterality and the growth of intellectual abilities. *British Journal of Educational Psychology 44,* 37–46.
Bradshaw JL & Nettleton NC (1983). *Human Cerebral Asymmetry.* New York: Appleton–Century–Croft.
Bradshaw JL, Nettleton NC, Taylor MJ & Mercelito J (1981). Right hemisphere language and cognitive deficit in sinistrals? *Neuropsychologia 19,* 113–132.
Bryden MP (1986). Dichotic listening performance, cognitive ability and cerebral organisation. *Canadian Journal of Psychology 40,* 445–456.
Burnett AA, Lane DM & Dratt LM (1982). Spatial ability and handedness. *Intelligence 6,* 57–68.
Charman DK (1980). Note on a failure to find hemispheric asymmetry for a small sample of strongly left-handed and right-handed males and females using verbal and visuospatial recall. *Perceptual & Motor Skills 51,* 139–145.
Corballis MC (1983). *Human Laterality.* New York: Academic Press.
Coren S (1992). *The Left-Hander Syndrome: The Cause and Consequence of Left-Handedness.* New York: Free Press.
I disagree with so much in this book, but the enthusiasm is infectious.
Crawford JR & Parker DM (1989). *Developments in Clinical & Experimental Neuropsychology.* New York: Plenum Press.
Elbert T & Birbaumer N (1987). *Individual Differences in Hemispheric Specialisation.* New York: Plenum Press.
Fennell E, Satz P, van den Abell T, Bowers D & Thomas R (1978). Visuospatial competency, handedness and cerebral dominance. *Brain & Language 5,* 206–214.
Goffman E (1968). *Asylums: Essays in the Social Situation of Mental Patients and Other Inmates.* Harmondsworth: Pelican.
Hardyck C & Petrinovich LF (1977). Left-handedness. *Psychological Bulletin 84,* 385-404.
Hardyck C, Petrinovich LF & Goldman RD (1976). Left-handedness and cognitive deficit. *Cortex 12,* 266–279.
Hermann DJ & Van Dyke KA (1978). Handedness and the mental rotation of perceived patterns. *Cortex 14,* 521–529.
Hicks RA & Beveridge R (1978). Handedness and intelligence. *Cortex 14,* 304–307.
Hicks RA & Dusek CM (1980). Handedness distributions of gifted and non-gifted children. *Cortex 16,* 479–481.

Hicks RA & Kinsbourne M (1978). Human handedness. In M Kinsbourne, *Asymmetrical Function of the Brain.* Cambridge: Cambridge University Press.
Jariabkhova K (1980). Pokazateli intellektualhogo razvitiya u pravoruchnykh i levoruchnykh dety [Intellectual achievements by right- and left-handed children]. *Studia Psychologica 22,* 249–253.
Johnson O & Harley C (1980). Handedness and sex differences in cognitive tests of brain laterality. *Cortex 16,* 73–82.
Kitterle FL (1991). *Cerebral Laterality: Theory and Research: The Toledo Symposium.* New York: Lawrence Erlbaum.
Kocel KM (1977). Cognitive abilities, handedness, familial sinistrality and sex. *Annals of the New York Academy of Science 299,* 233–241.
Kovac D, Jariabkova K & Zapotocna O (1984). Semilongitudinal study of laterality, cognition and personality. *Studia Psychologica 26,* 71–74.
Levy J (1969). Possible basis for the evolution of lateral specialisation of the human brain. *Nature 224,* 614–615.
McGlone J & Davidson W (1973). The relation between cerebral speech laterality and spatial ability with special reference to sex and hand preference. *Neuropsychologia 11,* 105–113.
McGlone J & Kertesz A (1973). Sex differences in cerebral processing of visuospatial tasks. *Cortex 9,* 313–320.
McKeever WF (1986). The effects of handedness, sex and androgyny on language laterality, and verbal and spatial ability. *Cortex 22(4).*
Miller E (1982). Handedness and a test of cognitive development. *Neuropsychologia 20,* 155–162.
Milner B (1972). Interhemispheric differences and psychological processes. *British Medical Bulletin 27,* 272–278.
An exciting model of logical clarity.
Milner B, Branch C & Rasmussen T (1966). Evidence for bilateral speech representation in some non-right-handers. *Transactions of the American Neurological Association 80,* 42–57.
Milstein V, Small IF, Malloy FW & Small JG (1979). Influence of sex and handedness on hemispheric functioning. *Cortex 15,* 439–449.
Peters M (1991). Reanalysis of Benbow's data on mathematical giftedness. *Canadian Journal of Psychology 45,* 415–419.
Porac C, Coren S & Duncan P (1980). Lateral preference in retardates: relationships between hand, eye, foot and ear preference. *International Journal of Clinical Neuropsychology 2,* 173–187.
Rose S, Lewontin RC & Kamin L (1990). *Not in our Genes: Biology, Ideology and Human Nature.* Harmondsworth: Penguin.
Sanders B, Wilson JR & Vandenburg SG (1982). Handedness and spatial intelligence. *Cortex 18,* 79–80.
Satz P (1973). Left-handedness and early brain insult: an explanation. *Neuropsychologia 11,* 115–117.
If every journal three pages were as good as this, we'd be a lot further forward.
Satz P & Fletcher JM (1987). Left-handedness and dyslexia: an old myth revisited. *Journal of Pediatric Psychology 12,* 291–298.
Smith ME (1986). Critique of E Miller "Handedness and a test of cognitive development". *Neuropsychologia 24,* 453–454.
Springer SP & Deutsch G (1989). *Left Brain, Right Brain, 3rd edn.* San Francisco: WH Freeman.

Swanson JM, Kinsbourne M & Horn JM (1980). Cognitive deficit and left-handedness: a cautionary note. In J Herron, *Neuropsychology of Left-Handedness.* New York: Academic Press.

Teng EL, Lee P-H, Yang K-S & Chang PC (1979). Lateral preferences for hand, foot and eye, and their lack of association with scholastic achievement, in 4143 Chinese. *Neuropsychologia 17,* 41–48.

Warrington EK, James M & Maciejewski C (1986). The WAIS as a lateralizing and localizing diagnostic instrument: a study of 656 patients with unilateral cerebral lesions. *Neuropsychologia 24,* 223–239.

Williams SM (1987c). Differences in academic performance at school depending on handedness: matter for neuropathology? *Journal of Genetic Psychology 148,* 469–478.

Wittenborn JR (1946). Correlates of handedness among college freshmen. *Journal of Educational Psychology 37,* 161–170.

CHAPTER 7

Are We Born or Made?

Endlösung.
Final Solution.

Virtually the whole of Psychology bears, or can be brought to bear, on the nature/nurture debate, so a comprehensive coverage of it is unrealistic. On the other hand, the title *Environment and Mental Health* demands some selective coverage, and that is what follows.

ENVIRONMENT AND POLITICS

It is often convenient to believe that our heredity destines us for a certain position in life. We are, thereby, absolved of responsibility for anything about it that we do not like. Furthermore, we find our place easier to tolerate. Accepting this idea can make it a reality, a "self-fulfilling prophecy". We do not believe we have any power to alter our station, and so make no effort to do so. The fact is, that nature/nurture is also politics. While people believe that their intelligence and general character are largely fixed at birth, they are more likely to accept their lot without complaint.

This point is unlikely to escape the committed conservative. It is probable, too, that conservatives have the greatest benefit from property and titles—that is, the greatest advantage from inheritance in the legal, non-biological sense.

It is comforting to justify such non-biological inheritance in terms of parallel genetic inheritance of such qualities as high ability. An American psychologist, Herrnstein (1973), has explicitly provided such justification, claiming outright that socio-economic status is a result of genes (asserting such status to be derivative from genetically transmitted IQ). I find such claims totally absurd in the current British context.

Other than class, the group differences that are most politically sensitive, where nativist positions have to be contended with, are those of race and gender. These are the areas where, in Britain, legislation has been required; the Sex Discrimination Act of 1975, and the Race Relations Acts of 1968 and 1976. As a white, middle-class male, I do not have the same kind of feelings about these sorts of issues that those more affected by them do. I totally oppose discrimination, and feel that beliefs or assumptions that group differences are immutable and genetic may often underpin it, in practice.

Since many have written books about these issues, perhaps I am unwise to show my ignorance by raising them at all. As my later discussion of "the Third Reich" will show, it is racism that concerns me most. There are racist organisations in Britain, such as the National Front and the British Movement, but they are weak. The more disturbing manifestation is le Pen's National Front in France (this may be because France is more exposed to North African demography). Such groups talk a great deal about "repatriation" and about restricting immigration. The idea of expelling anyone who has lived in the United Kingdom for some years is wholly abhorrent to me. Immigration is a more difficult issue, since there may be limits to the size of population a country can support, but again, to select immigrants on the grounds of their racial type is thoroughly immoral.

The United Nations Educational, Scientific and Cultural Organisation (UNESCO), in 1951, declared that, "Available scientific knowledge provides no basis for believing that the groups of mankind differ in their innate capacity for intellectual and emotional development". Though, at the time, this disregarded the dissent of a vocal school of academics, students of real genetics, at least, would today assent. Perhaps there will always be some who disagree with it, but I think they tend to disregard the relative rarity of pure racial types. Also, the effects of the racial type of those who attempt to appraise "intellectual and emotional development" contaminates much of the "evidence". In multi-cultural countries, publicists of psychological differences depending on racial type are playing with fire. Their assertions are socially irresponsible.

On gender, I find it troublesome to summarise my feelings in writing. I find it tempting to ascribe many of my own developmental problems to the social phenomenon called feminism, and to the critics of a "patriarchal" society. So, although I agree with thoughtful critique of the innateness of gender differences, such critique often seems to go with other ideas that I dislike. I recognise that there have been many healthy effects of the movement of women into new occupations. Some of these, it has to be admitted, are occupations for which they were once held not to be suitable. I should regard any suggestion that a woman is biologically unfitted for an occupation as wholly unwarranted. I think there is good evidence that women can make fine soldiers, for example. Yet I feel an anti-male quality about some feminist literature, as well as an ideological purism in a premise that the sexes are identical. At the same time, I think some men fail to recognise that the threats of rape and violence are more a part of the everyday environment for women. These are bound to have effects on their behaviour that might be mistakenly attributed to their biology.

I should like to say a little about homosexuality. Researchers have done twin studies of this. An early study by Kallman discovered 20 people, who were each one of an identical pair and were also, according to his criteria, homosexual (see Kallman, 1938). In every case, the twin partner turned out to be homosexual also. The result is too good to be true, for one thing, and other studies contradict it, anyway (Ruse, 1979: 133). This inconsistency suggests such studies are prone to much error, as well as logical problems that I shall describe. I believe there is a continuum of homosexuality, or, to use a cognate psychological term, androgyny, and so there is bound to be much unreliability about describing people as homosexual.

In this book, I have named some who deny any environmental influence on mind, some who deny any genetic influence. Such extreme positions have led all too readily to two separate camps, conducting a dialogue of the deaf. So often people talk past each other, when addressing a genuinely important social issue.

There was a past nature/nurture controversy regarding the aetiology (causes) of tuberculosis. Once, society rested content that the predisposition to tuberculosis (as to syphilis) was inherited. Thereby, people more easily ignored the horrific conditions of overcrowding in slums, which actually caused the high prevalence of the disease in Europe (up to 40 or 50 years ago). Tuberculosis did run in families, and twin evidence was also brought forward to support the hereditarian position. Thus Kallman found that if one twin has tuberculosis then concordance (the other twin also having the disease) was 3.5 times greater for monozygotic (that is, genetically identical) than for dizygotic (ordinary fraternal) twins. However, even within its own terms, this approach brought contradictions. Monozygotic twins living apart show no increased correlation with respect to tuberculosis, in line with what we now know to be true, that the higher concordance for monozygotic twins living together is a result of correlated environments. After too many years of Conservative rule in Britain, producing a new "underclass", tuberculosis seems on its way back.

The same readiness to believe in bad heredity seems to lie behind another old theory of genetic susceptibility: that some people inherited a vulnerability to the virus that causes hepatitis B. This, too, has turned out not to be substantiated by facts. The disease is commonest among the poor, among blacks, and in the tropics—due to their unsatisfactory living conditions.

As long as I can remember, I have believed that, after secure defence, the first duty of a government to its people is to remedy the effects of poverty. Politically, I am a Social Democrat, committed enough to have joined that party in Britain. Broader evidence suggests that there may be an entwining of political standpoint with nature/nurture viewpoint. A study of researchers themselves is described in Pastore (1949). This found that

11 out of 12 who defended the genetic standpoint on IQ belonged to the conservative party of their country, and almost all defenders of the environmental theory saw themselves as liberal or left-wing

Let me move to a historical perspective. Vroon (1980: 83) has said that the differences in opinion regarding heredity or environment are approximately the same as those of a century ago. This is a disappointing judgement, for one who believes that progress and modernity carry with them an increasing belief in environment. I should explain this myself in terms of cyclical swings, caused by the process of dialectic, between nature and nurture as underlying constructs, on top of a longer progressive trend.

The history of Psychology, itself, with respect to this issue, presents a confusing picture. Within behaviourism, starting with J.B. Watson, environmentalism has been prepotent. B.F. Skinner's own autobiography (Skinner, 1976) says almost nothing about his own personality, and presents himself as a creature of circumstance. To my eyes, it is the attempt to establish consistent individual differences that so often leads on to nativism. That project has been far more than a mere historical preliminary to behaviourism, though I should say the advance of the latter movement has thrown it into the shade. The current academic vogue called Cognitive Psychology includes great stress on Chomsky's psycholinguistics (see Chomsky, 1965), a nativist theory. Yet Chomsky, himself, is a noted American man of the Left (for example, author of *Why are we in Vietnam?*). This shows that to associate nativism with conservativism can, at best, be only a first approximation to the complexity of people's political beliefs (see Chomsky, 1992).

ENVIRONMENT AND THE ACADEMIC

I should contend that most people have thought about the nature/nurture issue in their own lives, perhaps in their own words and without realising they were doing so. Yet it is only by the public act of writing about it that someone can create a "name". Thereby she or he achieves a sort of eminence. Is eminence attained by innate qualities? This is not just one more aspect of the general issue, because those whom we read may be biased by the way they explain their own productivity.

It is not so long since the standard biography of a great man of science was a piece of piety that converted him into a "monster of perfection". F. Scott Fitzgerald wrote a story called *The Rich are Different*, and the view that the eminent are just "different" leads on to nativism.

Typically, eminent scientists are highly motivated, driving, indefatigable and prone to identify with their work. It seems likely that prestige and recognition are what motivate them. In a 1963 study of the rank order of ninety occupations, the most highly rated scientific occupation—nuclear physicist—held third place. Other scientific occupations, too, were rising in the rank order.

At the highest level, it is the drive for the Nobel prize that contributes much motivation. Zuckerman (1977) wrote a book containing much interesting material on the psychology of Nobel laureates. She showed how they apprentice themselves to eminent scientific masters, often laureates already, and will tolerate ambivalence and conflict in their relationships with them. These "friends of promise" (to invert Cyril Connolly's *Enemies of Promise*) tend to be promoted early. The great pure mathematician, G.H. Hardy, said the motivation is all ambition, curiosity and pride: "If a mathematician, or a chemist, or even a physiologist, were to tell me that the driving force in his work had been the desire to benefit humanity, then I should not believe him". One sign of fierce ambition is the prevalence of highly competitive practices, such as keeping one's ideas secret until they are ready to be publicly unveiled. This was very vividly portrayed in James Watson's *The Double Helix*, about the discovery of the chemical structure of DNA (his collaborator, Francis Crick, has written a more recent autobiographical book, *What Mad Pursuit*).

It seems obvious to me that early experience, perhaps of parental neglect, can explain needs for recognition. Or the confidence to drive for recognition can be engendered by early parental encouragement.

According to Zuckerman, various regularities appear in the background of Nobel laureates. They take their PhDs from an élite group of institutions. Their early productivity is promptly recognised, and is transformed into resources for further work. Less obviously, as a group they have a distinctive family religious background. Jews are over-represented nine times in relation to the US population, Roman Catholics under-represented 25 times. Religion plays little part in the lives of the scientists themselves, however.

Roe (1953) made further suggestions about the backgrounds associated with high productivity. The scientists in her sample married in their late twenties, that is, rather late; partly because of financial considerations, but also because of preoccupation with their work. The financial pressure on those who are not, like this and like most samples of the eminent, predominantly from high socio-economic-status families, may of course be crippling in terms of intellectual productivity. It should be noted, however, that although professional backgrounds afford less relief from financial constraints than do business ones, these are the ones from which the scientifically eminent tend to come.

Well over half Roe's sample, many more than would have been expected by chance, were first-borns. Other evidence suggests that not only birth order but also the unexpected death of a sibling or parent (as well as separation from one, or "diminution in her saliency") are factors in later eminence. Children who are very close to their parents, and to adults generally, rather than to other children, may be destined for later success.

The most important characteristic of the home background of Roe's sample was that "learning was valued for its own sake". Terman (1925), too, with his classic follow-through of high-IQ children, found that a large family library was a typical feature (as was breast-feeding!). Roe also suggests that the boy who cannot, for some reason, such as physical disability, compete effectively in sports, can gain status in another way, by surpassing other boys in school work. I can speculate to what extent this idea applies to girls as well. There is also the factor of a period of physical illness. Though it obviously impairs productivity at the time, Roe says that it can often act as a watershed period, in which the future person of ideas becomes decisively oriented towards intellectual pursuits.

Other important background factors, thinks Fox (1983), are early autonomy, independence and self-sufficiency. There is evidence that creativity can be hampered by excessive discipline, and even by too much formal education. Creative thinkers prefer teachers who let them alone, and like to feel they have "internal locus of control".

In a similar vein, Ben-David & Sullivan (1975) say, that too co-ordinated an organisation can stultify the creativity of its members. There is more work besides on the effects of the organisational context. When a scientist moves to a prestigious institution, particularly one with high collegiality, his productivity increases (Fox, 1983).

The *age* of Nobel laureates, when they do their prize-winning research, strongly suggests that life experience is an important foundation for the peak act of intellectual production (Zuckerman, 1977). The science prizes are only awarded to living scientists. Because of the secular increase in the number of scientific jobs, most scientists are young. Yet the American laureates Zuckerman interviewed were 40, on average, when they did the research. Most of them were productive over long time-spans of 30 years or so. Most of the Nobel research is collaborative, and, so, the importance of social dynamics is obvious.

The productivity of laureates declines sharply in the 5 years following receipt of the prize, which proves the role of environmental factors. The decline is particularly sharp among those relatively less eminent before receipt of the prize. The social system induces a sudden outbreak of requests for advice, speeches, review articles, and greater participation in policy decisions and other public services, which distracts from scientific production. Nobel laureate Davisson said, "I was transformed overnight from an exceedingly private person to something in the nature of a semi-public institution" (see Zuckerman, 1977). An additional reason for the decline is greater hesitancy among initiated Nobel laureates to publish any work that might be judged mediocre.

Simonton (1978) has broadened out the idea of environment to consider the influence on the eminent person of his general historical context. Simonton describes various historical factors. These include the nature of formal education at the time, the availability of role models, the *Zeitgeist*, and whether there is political fragmentation, or war, civil disturbance or general political instability. All these, he says, tend to have a critical impact on the development of creative potential in the young genius. On the other hand, he says that once that potential is established, and the genius enters adulthood, creative productivity proceeds with little influence from outside events. In Simonton's view, the most famous thinkers are synthesisers, who take the accomplishments of the preceding generation, flowing from that previous *Zeitgeist*, and consolidate them into a single unified philosophical system.

I should say, then, that there are plenty of ways in which we can understand the exceptional intellectual productivity of certain people, without simply thinking of them as "different". For as long as they think of themselves as different, mystique of native abilities will persist in expert discourse.

THE THIRD REICH

To show how the nature/nurture issue matters to us all, the obvious case is the "Third Reich (empire)". This is the most dramatic example so far of the horrific power of extreme nativist theories over a whole country. The Führer (leader) of Nazi Germany, Adolf Hitler, accepted completely an idea of hereditary taint (*erblich belastet*), with dreadful consequences.

The State authorities required every German family to provide a very detailed genealogy. The experts who drew these up became VIPs, for discovery in the family tree of an individual from a stigmatised group, above all of a Jew, implied a degree of taint. The value of this, in a tyrannical regime based on fear, was, of course, that every family had something to hide, if it went back far enough, or thoroughly enough. Germans were deemed to have tainted heredity if one parent, or two siblings, or a third of other relatives suffered from either severe physical or severe mental abnormality. Suppose two quite different conditions, such as schizophrenia, in one relative, and a cleft palate, in another, appeared in the family. The sufferers were added up toward the number needed to diagnose hereditary taint. Homosexuals and gypsies were also deemed tainted.

Hitler believed Gobineau's (1853) theory, that the reason why empires through history had declined was that the originating race had lost its "fitness" by out-marrying. This was a misapplication of the Darwinian idea of fitness, which means "being adapted to the environment". So, one basis of Hitler's beliefs was a crude Social Darwinism (Bullock, 1991: 157). He called the founding, pure-blooded German race (Nordics, blue-eyed and fair-haired), "Aryans". This was the *Herrenvolk* (master race), who, while they kept themselves pure by marrying within themselves, incurred no hereditary taint. Appointments in the gift of the state went only to untainted Aryans. Hitler intended them, and them alone, to be the future of Germany: only they received state marriage loans, that is, only couples approved as "Aryan". Certain types of marriage (i.e. those involving tainted individuals) were forbidden by law. Those with hereditary disorders were allowed to marry only someone who was sterile. The sterile were not to lessen the fertility of an Aryan, and were not allowed to marry unless it was to another who was sterile.

One natural consequence of the Nazi tainted heredity idea was an emphasis on artificial sterilisation of those so tainted, and indeed, under the *Erbgesundheitsrecht* (congenital disease law), compulsory sterilisation. By 1939, 375 000 people had been sterilised under the complete programme, most for "congenital feeblemindedness", but 4000 for blindness or deafness (Gould, 1985: 309).

Bullock (1991: 157) describes Hitler's *Mein Kampf* (*My Struggle*) as "a book which has few rivals in the repulsiveness of its language, its tone and above all its contents". He thinks (p459ff) the racism set forth in it was Hitler's personal contribution to Nazi ideology, making that ideology distinctive. Surely, it emanated, at least in part, from some traumatising personal experience involving a Jew (see p160). So his view is that racism is not necessarily a standard feature of Fascist movements. The Nazis were, of course, the National Socialist party. Bullock sees their racism as an extreme form of nationalism, flowing from the historical strategic German policy for *Lebensraum* (living space) to the East.

The concluding episode of the British television documentary series *The World at War* is entitled *Remember*. Though Auschwitz is behind us, the world must remember the terrible consequences of becoming gripped by this kind of madness. Remember Dachau and Belsen, remember Buchenwald, remember the horrors to which "purification" of the human family leads.

SOCIAL BIOLOGY AND EUGENICS

"Eugenics" is the use of genetic or pseudo-genetic principles to improve the quality of the human "stock". Thus "positive eugenics" is encouraging the "fit" to have children. "Negative eugenics" is discouraging the "unfit" from having children, which is obviously in a different universe ethically. Such ideas have always been controversial. The American Eugenics Society re-emerged in 1972 as the Society for the Study of Social Biology, perhaps because they felt the Left were too successfully turning "eugenics" into a hate-word. There is, of course, a huge difference between an actual eugenic programme and academic study.

It should not be thought that the *Erbgesundheitsrecht* was just a local, historical aberration. The Nazi programme was just an extreme version of an ever-present social tendency.

In 1974, in the USA, Judge Gesell came to a grave conclusion. He said that the threat that various federally supported welfare benefits would be withdrawn has improperly coerced an indefinite number of poor people in his country into accepting a sterilisation operation. Coming even more up to date, the notorious American eugenicist William Shockley (a Nobel laureate) has talked about sterilisation of people with "bad heredity". With true academic caution, he proposed this as no more than a "thinking exercise". The social menace of compulsory sterilisation does not automatically recede with the progress of civilisation. Technically, it has been eased by relatively safe and simple operations such as vasectomy and salpingectomy (cutting or tying of Fallopian tubes), to replace castration and other forms of mutilation now socially unacceptable.

The eugenics movement was launched by Francis Galton, the biography of whom by Derek Forrest (1974) is subtitled "a Victorian genius". *Erblich belastet* was, really, an extension of a British nineteenth-century idea that followed through some insinuations of the work of Charles Darwin.

Charles Darwin's theory of the evolution of biological species was a major scientific advance, though "creationists" and "creation science" still deny its truth, even today. Darwin said that all species, including our own, arise from a very long and slow process, of evolution through "natural selection". In a world with limited resources, more animals are born than can survive and reproduce. So, there is a struggle, in which the "fitter", including their heritable features, are "selected". It is in this way that there emerge "adaptive" organs, like hands and brain neo-cortex. In Victorian times, Darwin's theory broke on some as perniciously irreligious, on others, such as Galton, as wonderful enlightenment.

Its extension by another Victorian, Herbert Spencer, to what could be called the evolution of human society, was only a matter of time. It is intriguing to note that Spencer himself was always a follower of Lamarck. This Frenchman believed what biologists now regard as heresy: that characteristics acquired during the lifetime of one generation can be passed on genetically to the next. In our own time, Arthur Koestler has persuasively refused to let this belief die.

Of course, most social changes cannot have been caused by genes. The rise and decline of Islam took less than 30 generations, and so cannot have been driven directly by the genes. A modern epigram is that "We are raised according to the findings of child psychologists, and raise our children following the newer findings of child psychologists" (Leahey, 1987: 479). This is an example of the phenomenon of social evolution: change can occur by enabling the direct cultural transmission of innovative behaviour and technology across generations and between groups in a non-genetic fashion. None the less, ideas of fitness and selection can readily be adapted to social phenomena. I find it unlikely that evolutionary theory should have nothing to say to us today, in spite of the terrible aberration, in Germany, that I have described. In 1975 E.O. Wilson wrote *Sociobiology: the New Synthesis*, which has been a landmark in the modern reconsideration of these issues.

Wilson, a world authority on the social insects, convinced many that evolutionary ideas can be extended from the explanation of physical organs to the understanding of social behaviour. What does it mean to speak of "social behaviour" in the insects? One well-known example (work of von Frisch, 1966) is that scouting bees can return to the hive, and communicate to the residents, by means of a coded dance, the location of some pollen they have found.

Wilson went on to argue that such behaviours as aggression, attraction, and altruism are to be understood in hereditarian and evolutionary terms.

Another exponent of this sort of view was Konrad Lorenz (1966), who wrote a book *On Aggression*. He grappled with the evidence that in the animal kingdom fights within a species ("intraspecific" fights) are not like prey–predator aggression. They are restrained, involving ritual, bluff, non-fatal violence and appeasement gestures. This seems to argue against the comparison with human aggression. Lorenz's view was that natural selection has somehow gone wrong with us, the human species. This strikes me as a very *ad hoc* adjustment to try to rescue the evolutionary comparison, and perhaps a reflection of despair following the two World Wars. Wilson takes a different tack, claiming that in the insect world, at least, there certainly are intraspecific fights to the death.

With regard to attraction, I do find it worth at least pondering whether, for example, sibling incest taboos, in human cultures, have parallels in animal species (they do: Ruse, 1989: 172–178).

Human altruism poses a particular challenge to sociobiologists. There is something in the metaphor of a struggle for survival that is uncomfortable with the idea of unselfishness. But they have certainly not shirked the task, and there is plenty of work on the adaptive advantage of mutual helping. There is also the idea that human evolution has been about the survival of groups, rather than of individuals, with the "fit" group being one within which unselfish individual helping occurs.

Yet, surely, to extend ideas from the animal realm into that of human beings is a huge leap; above all, perhaps, with respect to human thinking.

One other form of social behaviour to which hereditarians have turned their attention is crime. The life and culture of criminals have often puzzled biologists and psychologists, closed into their scholarly lives, who have looked to heredity for the answer.

The early representative of the approach, and of its deficiencies, was Lombroso (1895). He compared the bodily appearance of prisoners with that of soldiers (hardly a purely random collection of physical specimens). Unsurprisingly, when he counted what he called bodily "anomalies", the prisoners had more, and these he imputed to be genetic. Already, by 1913, a careful attempt to repeat this study, using, for comparison, not men it had been judged could fight for their country, but a random sample of non-prisoners, did not find the same result. Also, it is plain enough that a bodily anomaly, perhaps causing a deformed appearance, may drive the person in question into social isolation, and into crime. So even if a correlation were found, it would not show anything genetic.

Is Hans Eysenck a modern Lombroso? He believes people with an extroverted personality have inherited a nervous system that is less easily conditioned. Since he also believes extroverts are more likely to become criminal, he, like Lombroso, believes there is a congenital factor in criminality.

It is easy to be misled, as I have said before and shall shortly repeat, by evidence that something, such as crime, runs in families. One old study presented a family called Jukes as an example of "a family that had degenerated because of heavy inbreeding" (see Vernon, 1979). The researcher said he had traced 2000 descendants of the original Jukes immigrants to the USA. Of these, 378 died in infancy, 301 were born illegitimate, 366 were paupers, 80 were habitual thieves, 171 were convicted of other crimes including 10 murders, 175 were prostitutes, and 50 were known to have venereal disease. This record could be contrasted with that of the Edwards family. Of 1394 descendants of Jonathan Edwards, 15 became presidents of universities, 65 were professors, 60 were doctors, 100 joined the clergy, 75 were army or navy officers, 60 were prominent authors, 130 were lawyers (30 of them judges), 80 were public officers and three were Senators. The two families could be "before and after" in a television commercial! This was the sort of evidence that underlay geneticist views of crime. Nowadays, such contrasts are more readily regarded as constellations of social factors rather than as introducing our sinister acquaintance, "tainted heredity".

I think it is undeniable that, in a mild way, eugenics has always had an influence. Thus, the British Home Secretary, not so long ago, refused the wife of a man serving 18 years in gaol artificial insemination by donor from her husband. When such specific cases arise, eugenic preconceptions may have an influence.

There is a very frightening potential avenue for an influence of eugenics.

This is the perversion of the perfectly legitimate and valuable genetic counselling, that takes place for the control of acknowledged genetic diseases (such as cystic fibrosis).

Genetic counselling influences people greatly. In one Edinburgh study, of all those told that they were at high (defined as more than one in ten) risk of having a congenitally impaired child, four out of five decided to avoid a pregnancy. Almost all of them carried out the decision successfully.

Vigilance will always have to be maintained to see that pseudo-science about the genetic causation of, for example, IQ and mental illness is not abused. People who have suffered mental disturbance, or who score low on IQ tests (often, quite simply, the poor) should not be prevented from having children. The public must be made aware of the distinction between Mendelian single-gene disorders, and so-called "polygenic disorders".

KINSHIP STUDIES

Nativists have, none the less, provided a form of evidence for their doctrine, in the shape of kinship studies of families, including those with adopted members, and of twins. This field is known as "behaviour genetics". The purpose of the following section is to show that these studies have major weaknesses as evidence for genetic determination.

Medawar & Medawar (1977) have said: "Geneticism is a word that has been coined to describe the enthusiastic misapplication of not fully understood genetic principles in situations where they do not apply. IQ psychologists are among its most advanced practitioners." Another leading scholar, Jencks (1973), has said: "our main conclusion after some years of work on this problem is that mathematical estimates of heritability tell us almost nothing about anything important".

True, as opposed to "behaviour", geneticists tend to regard the latter's efforts as theoretically simplistic. For example, biological geneticists work with concepts known as "dominance" and "assortative mating". Dominance and assortative mating have opposite effects on phenotypic variation within families as opposed to between families and cancel each other out. This is fortunate, since very few analyses of heritability take either factor explicitly into account. IQ scores of human mates do correlate above 0.3 (this is the phenomenon of assortative mating), which will reduce IQ variability within a family and so artificially inflate the estimate of IQ heritability.

The central thesis of Gottesman's (1992) *Schizophrenia Genesis* is that nature is more important than nurture for this diagnosis. At one stage the author "attributes the fundamental causes of schizophrenia to genetic factors", because, though "schizophrenia is both a genetic and an environmental disorder, [there is a] relatively low prevalence of the genetic predisposition compared with the relatively high prevalence of the various environmental causes/triggers/risk-factors" (p218). Elsewhere, he presents "an ecumenical [diathesis–stressor] model that permits many flowers to bloom, and cannot be refuted" [!] (p231). Where he is clear is that first-degree relatives of schizophrenics do not show 50/25% rates of the disease expected for a Mendelian dominant/recessive single-gene disorder. He favours a polygenic model for which "family, twin and adoption studies provide the grist for our mill. Each contributes to the genetic argument, complementing the others. No one method alone yields conclusive proof or disproof."

It should be obvious that the fact that something runs in families does not prove genetic determination. There is plenty of evidence that mental disturbance does run in families (a recent interesting review is Romney, 1990), but there are other ways of looking at this. Laing (1967) put this point in a neat reversal, "The syndrome of psychiatry runs in families—it is a pathological process called 'psychiatrosis' with psychic correlates, an inherited basis, a natural history and a doubtful prognosis". We could explain evidence that mental disturbance is familial in terms of disturbed parents providing an insufficiently stable early environment. We could explain evidence that intelligence is familial in terms of intelligent parents providing a stimulating early environment.

There are not only logical points of this type. What is the quality of the data? To me, human kinship samples seem ragged in the extreme, compared to the factorial experimental designs I cut my own teeth on, and found, themselves, woefully deficient in yielding any firm conclusions. They often do not even have equal numbers of subjects in each group being studied.

In no way can the inferences from data about twins to heritabilities be taken as simply mathematical and guaranteed. The average IQ for all twins studied is only 95, though average IQ for the general population is defined to be a hundred. Since twins are not typical, it is unwise to place on evidence from them the reliance that has been placed.

Siblings and fraternal twins of schizophrenics show about a 12% incidence of the disorder, while identical twins show a 40–50% incidence (Ruse, 1979: 131). The assumption underlying inferences about heritability is that the environments of identical and fraternal twins are equally similar, and so any differences between identical and fraternal concordances must be due to genetic factors. This is wrong. For example, parents, teachers and peers may treat identical twins more similarly than they treat fraternal twins. Since they often dress the same, with the same hair styles and so on, they are often mistaken for each other, with all sorts of social consequences. It is known for identical twins to create a personal language in which to talk with each other. Yet Gottesman can speak in his book of their "alleged" identity confusion.

Even the identification of twins as "identical" can be done on different bases: either fingerprints, or eye colour, or hair colour, or height, and so on, i.e., it is not infallible. Torrey (1990) has criticised the evidence from monozygotic twins in terms of geneticists applying the criteria for schizophrenia too loosely when it suits them. Gottesman recognised this problem, and tried in one study a narrow definition of schizophrenia: the small sample showed higher *dizygotic* than monozygotic concordance.

The "classical studies" in this vein, by investigators like Kallman, were clearly contaminated by personal hereditarian bias. Rüdin (see Gottesman, 1992) investigated the genetic aspects of the discovery of "schizophrenia" by his collaborator Kraepelin (1887). He became thoroughly convinced that environmental factors were also critical in determining who became schizophrenic. But who of these early sole investigators paid any regard to the need for blind determination of diagnosis or zygosity?

Twin concordances have been moving results ever since. Early studies of schizophrenia suggested a concordance of 70% for identical twins, but this type of result was obtained for schizophrenic twins discovered through hospital lists. Surveys of the general population give lower estimates, around 50%. Surely the fact that there is any discordance at all shows that genes are not the whole story.

Heritabilities are also estimated from studies of identical twins reared apart, with the assumption that psychological similarities must reflect genetic duplication. Yet adoption agencies ensure that the environments of separated twins are often highly correlated. Vernon (1979) rejects some of Kamin's criticisms of separated monozygotic twin data. But he admits that separated monozygotic pairs do tend to be placed in similar homes, which would tend to boost their IQ correlations, and so the heritability estimate.

Gottesman acknowledges (1992: 107) that "Small sample sizes are the rule in twin studies of psychopathology". This seems like an understatement when it comes to separated monozygotic twins. Ruse (1979) says that only 17 pairs of identical twins, one or both of whom are schizophrenic, and who were separated near birth, are known in the world. Similarly, we find that Gottesman's assertion that the individual who has two schizophrenic parents has a lifetime risk for developing schizophrenia of 46% (p96), is based on 62 schizophrenics (p101). There is a serious mismatch between the gravity of the conclusions and the size of the sample.

Some studies look at remarriages, and compare correlations for siblings with those for half-siblings. But siblings and half-siblings differ in the way they are treated, as well as in their degree of genetic variation. There are similar studies of foster children. Munsinger (1975) is the psychologist who has provided the best discussion of the serious problems in collecting unbiased data on the intelligence of foster children relative to true and foster parents. Some studies since his review have found no difference between true and foster parents in the degree of association with the child's intelligence.

An adoption study by Kety *et al.* (1978) is one piece of direct evidence for a genetic component in schizophrenia. It is cited more than any other geneticist's study—Gottesman (1992) for example gives it several pages. Kety looked at children (as we have seen, sometimes a difficult group to diagnose), with schizophrenia, who had been adopted. He found a raised incidence of schizophrenia in their natural but not in their adoptive parents. But Kety found only 33 schizophrenics out of 5483 adoptees. His definition of "schizophrenia" included "latent schizophrenia", a vague construction that, without blind diagnosis, could easily lead a committed geneticist to see an illusory pattern. It is likely, that schizophrenia is more common among adults who are forced, probably by poverty, to relinquish their children, than among adults who are comfortable enough to convince an adoption agency that they are fit to adopt children. Finally, and directly contrary to the genetic view, a closer reading of Kety's research shows that he found more schizophrenics among the half-siblings of the child schizophrenics than among the true siblings or their parents. When a study so flawed as this one comes to take on such an important role in the geneticists' case one could be forgiven for supposing that case to be a weak one.

NEUROMYTHOLOGY

Protagonists of the hereditarian standpoint often have recourse to explanations in terms of the brain to provide the mechanism. One sort of physical mechanism, of genetic transmission through DNA replication, goes naturally with the physical mechanism of nervous tissue.

Thus it is sometimes said that girls develop language ability earlier than boys do. Conversely, they are also said to develop scientific abilities later (thus, there is not a single living American woman Nobel laureate). A 1972 theory of Tony Buffery and Jeffrey Gray suggests that these are innate differences, because a girl's left hemisphere matures earlier than a boy's. Fairweather (1982) on the other hand attacks this sort of evidence for sex differences in lateral asymmetry.

He holds to the "null hypothesis" of no difference. Henrik Ibsen said that the minority is always right, and so, in my view, is the null hypothesis. Academics have to publish, for the sake of their careers, and it remains difficult to publish a result of the form "I failed to reject the null hypothesis". "Publish or perish" should be annotated "so reject it or exit". I have written a chapter on this in my book *Psychology on the Couch*.

Incidentally, there is evidence (Pasewark *et al.*, 1975), suggesting that the "publish or perish" syndrome influences women academics less than men, no doubt partly because of other pressures towards child-rearing and home-making.

So I find little difficulty in believing Fairweather, rather than Buffery and Gray. It is wrong to pretend that only differences count. In the end, assertions in the older textbooks that women like monotonous jobs will be replaced or embellished by explanations that they have "skew-whiff" brains. Researchers in this tradition will cry "*Vive la différence*" with a second meaning. Statistically "significant" differences that carry no warning of their utter insignificance in size or reliability will continue to lead us down the garden path. Published "findings" will remain immune from contradiction by unpublishable negative data, antedating them possibly by years.

"Sex differences in hemispheric asymmetry" are just the tip of an iceberg. There ought to be terms, parallel to "creative accounting", like "creative neurologising" and "creative hereditising", so widespread is the practice of stretching evidence to fit an irrelevant neurological mould. For example, again in the lateral asymmetry field, 20 years of research on ear differences after Kimura (1961) did not even mention the fact that people put a telephone receiver to one or other ear (Surwillo, 1981; Williams, 1982). Most researchers thought only brain hemisphere differences relevant. It seems that many seek to justify and fund scientific curiosity by straining to assert links with physical structures with obvious medical importance, such as the brain. Hereditarianism flourishes above all because it supports the status quo; but the academic servants of hereditarianism flourish by encouraging scientific mystique.

ANNOTATED CHAPTER BIBLIOGRAPHY

Ben-David J & Sullivan TA (1975). Productivity and academic organisation in nineteenth-century medicine. In B Barber & W Hirsch, *The Sociology of Science*. New York: Free Press.

Buffery AWH & Gray JA (1972). Sex differences in the development of spatial and linguistic skills. In C Ounsted & DC Taylor, *Gender Differences: their Ontogeny and Significance*. Edinburgh: Churchill Livingstone.

This book seems to have been supplanted by one by Maccoby and Jacklin. The article seems untypical to me of hemispheric specialisation literature.

Bullock A (1991). *Hitler and Stalin: Parallel Lives*. London: Harper Collins.

A huge work by the historian who wrote *Hitler: A Study in Tyranny*, focusing on these men's public lives, to a large extent intertwined. Stalin was the older, and lasted longer. As Bullock insisted in his earlier book, "no man becomes a tyrant in order to keep out the cold". I see one of the tasks of our age to shake off the fear of individuals that so often means, at root, one of these two.

Chomsky N (1965). *Aspects of the Theory of Syntax*. Cambridge, MA: MIT Press.

Chomsky N (1992). *Chronicles of Dissent.* New York: AK Press.

Connolly C (1988). *Enemies of Promise.* London: Deutsch.

Crick F (1990). *What Mad Pursuit: A Personal View of Scientific Discovery.* Harmondsworth: Penguin.

I think Crick was right to wait so much longer than Watson to give an account of their epoch-making discovery.

Dobzhansky T (1962). *Mankind Evolving.* London: Yale University Press.

This includes a good deal of material on nature/nurture. It is written from the perspective of a biologist who can make his content consistently interesting and readable.

Fairweather H (1982). Sex differences. In JG Beaumont, *Divided Visual Field Studies of Cerebral Organisation.* London: Academic Press.

This was slated in a review by Professor Andrew Ellis.

Forrest D (1974). *Francis Galton: The life and work of a Victorian genius.* London: Paul Elek.

Fox MF (1983). Publication productivity among scientists: a critical review. *Social Studies of Science 13,* 285–305.

A very full and interesting source.

Frisch K von (1966). *Dancing Bees: An Account of the Life and Senses of the Honey Bee.* London: Methuen.

Gobineau JA, Comte de (1853). *Essay on the Inequality of Human Races.*

Gottesman II (1992). *Schizophrenia Genesis.* San Francisco: WH Freeman.

It takes a good writer to produce a thoroughly annoying book.

Gould SJ (1985). *The Flamingo's Smile: Reflections in Natural History.* London: WW Norton.

Herrnstein RJ (1973). *IQ in the Meritocracy.* London: Allen Lane.

This is so outrageous that words fail me.

Jencks C (1973). *Inequality.* Harmondsworth: Penguin.

I intend to read this myself as soon as possible.

Kallman FJ (1938). *The Genetics of Schizophrenia.* New York: Augustin.

Looking at the date again, I feel I may have been unfair to Kallman.

Kety SS, Rosenthal D, Wender PH, Schulsinger F & Jacobsen B (1978). The biological and adoptive families of adopted individuals who become schizophrenic. In LC Wynne, RL Cromwell & S Matthysse, *The Nature of Schizophrenia.* New York: John Wiley.

Kimura D (1961). Cerebral dominance and the perception of verbal stimuli. *Canadian Journal of Psychology 15,* 166–171.

A "citation classic".

Kraepelin E (1887). *Psychiatrie, 2nd edn.* Leipzig: Abel.

Laing RD (1967). *The Politics of Experience.* Harmondsworth: Penguin

Leahey TH (1987). *A History of Psychology, 2nd edn.* Englewood Cliffs, NJ: Prentice-Hall.

Lombroso C (1895). *L'homme criminel, 2nd edn.* Paris: Ancienne Librairie Germer Baillihre et Cie.

Lorenz K (1966). *On Aggression.* London: Methuen.

Medawar P & Medawar J (1977). *The Life Sciences.* New York: Harper & Row.

Munsinger, H (1975). The adopted child's IQ: a critical review. *Psychological Bulletin 82,* 623–659.

This journal is one of the two major sources for landmark Psychology review articles.

Pasewark RA, Fitzgerald BJ & Sawyer RN (1975). Psychology of the scientist: XXXII. God at the synapse: research activities of clinical, experimental and physiological psychologists. *Psychological Reports 36*, 671–674.
This journal has a high acceptance rate for submitted manuscripts.
Pastore N (1949). *The Nature–Nurture Controversy.* New York: King's Crown.
Roe A (1953). *The Making of a Scientist.* Westport, CT: Greenwood Press.
Romney DM (1990). Thought disorder in the relatives of schizophrenics: a meta-analytic review of selected published studies. *Journal of Nervous & Mental Disease 178*, 481–486.
Rose S, Lewontin RC & Kamin L (1990). *Not in our Genes: Biology, Ideology and Human Nature.* Harmondsworth: Penguin.
According to Gottesman (1992), "a politicized tract on the periphery of the authors' expertise". I, on the other hand, find even the historical theory rather interesting.
Ruse M (1979). *Sociobiology: Sense or Nonsense?* Dordrecht: Reidel.
A philosopher makes these biological ideas accessible.
Ruse M (1989). *The Darwinian Paradigm: Essays on its History, Philosophy, and Religious Implications.* London: Routledge.
An even better book than the one above.
Simonton DK (1978). History and the eminent person. *Gifted Child Quarterly 22*, 187–195.
There is a recent fascinating book on scientific productivity by this author (cited in the bibliography).
Skinner BF (1976). *Particulars of my Life.* New York: Knopf.
Another autobiographical book is *Cumulative Record.*
Surwillo WW (1981). Ear asymmetry in telephone listening behavior. *Cortex 17*, 625–632.
Terman LM (1925). *Mental and Physical Traits of a thousand Gifted Children. Genetic studies of Genius.* Stanford, CA: Stanford University Press.
Torrey EF (1990). Offspring of twins with schizophrenia. *Archives of General Psychiatry 47*, 976–977.
Vernon PE (1979). *Intelligence: Heredity and Environment.* San Francisco: WH Freeman.
Vroon PA (1980). *Intelligence: on Myths and Measurement.* Oxford: North Holland.
Watson JD (1980). *The Double Helix.* London: WW Norton.
A book that created a stir as much, sadly, for its revelations about the author's reaction to Cambridge, as for its account of his work.
Williams S (1982). Dichotic lateral asymmetry: the effects of grammatical structure and telephone usage. *Neuropsychologia 20*, 457–464.
Wilson EO (1975). *Sociobiology: the New Synthesis.* Cambridge, MA: Harvard University Press.
A landmark, but better for the general reader, before this, is Ruse.
Zuckerman H (1977). *Scientific Elite: Nobel Laureates in the United States.* New York: Free Press.
How she got so many of them to talk beats me.

CHAPTER 8

Summary and Conclusion

Better is the end of a thing than the beginning thereof:
and the patient in spirit than the proud in spirit.
Ecclesiastes 7:8

WHY SCIENTIFIC PSYCHOLOGY HAS NEGLECTED ENVIRONMENT

The important thing to understand about Psychology is that great energy is being invested, all the time, into convincing anyone who will listen, that it is a science. In the history of the universities, the subject achieved autonomy as a project for the *scientific* study of mind. To make that project more plausible, attempts have constantly been made to limit the scope of the subject, to draw it away from those areas to which scientific methodology seems less applicable.

At the same time, science is presented as a trans-human path to the truth. The effect of this, for many laypeople, is to imbue Psychology with great authority. In my view, that authority is wholly specious. We have to divest ourselves of any notion that science is trans-human. People produce it, and those people are similar in most important respects to anyone else. They are supported by a wider society—other people—and need, as much as anyone, to present themselves to others as worthy of support. I believe that it is only by bringing the whole range of academic qualities to bear on psychological issues that psychologists may hope to command attention.

Making an effort to see science in an historical perspective draws us towards seeing its human character. The rise of science was entwined with the rise of certain Western nations, principally, I should say, France, Britain, Germany and the USA. As, in our own time, the world hegemony of the West lessens, so may that of science or, at least, that of the Western mould of science. Origins in these particular countries meant that early science was shaped by its relationship with the Christian church. Though many would emphasise the tension in that relationship, I should say, myself, that to a large extent it was a constructive interplay. It is common to say that we now live in a "secular society", where this relationship, and what went with it, are less fundamental. One effect of this has been to open areas of life that were once the preserve of the Church to secular and scientific observation. The explanation of mental aberrations, for example, no longer needs to stop at a concept of "original sin".

Within the Church, it seems that it was particularly Nonconformists (the name Protestant conveys this quality of rebellion) who were more open to the ascent of a secular viewpoint on the world. This is in line with evidence that nonconformism, in the wide sense of a refusal to conform to authority, is helpful to the advance of disciplined study into new frontiers. I take the psychological effects of the environment to be such a new frontier.

Since, in Britain at least, authority generally derives from the past, defiance of authority means defiance of the past or, in other words, commitment to progress. I feel that part of what marks a branch of study as scientific is a "progressive dynamic". This is much of what gives it a claim to be taken seriously. What is very important, is to see the present as "a place along the path", and not as the end of the path. To continue the metaphor, the past is not merely the path so far, but also the country it leads through, where lies hidden much of relevance to the future of the path. Nor, of course, does the path lead straight. Not all change is progress, and opposing forces will play themselves out to a stable vector before the best path is found. This, to me, is all that is meant by the "dialectic". The picture of science in history as a play of opposing forces, backed as much by rhetoric as by reason, is being taken much more seriously in historiography. So that *controversy* is not a derangement within science, but rather is intrinsic to scientific progress. It is too simple to see scientific agreement as based upon agreed public facts. So many facts can only be checked by an elite, and it is the desire to be admitted to a club that motivates many who apprentice themselves to science.

Some controversy is healthy and necessary, but it also flourishes amid confusion. The environment is complex, and too much complexity breeds confusion. One sign of confusion is jargon, and I have to apologise for the amount of jargon that has forced itself into this book, particularly the piece of jargon coming up. It is because of its complexity, I should say, that the environment has fallen within the "negative heuristic" of Psychology, that is, the set of problems that that research programme has chosen to ignore. Yet a progressive dynamic should whittle away at the negative heuristic. This is probably more workable where we restrict ourselves to the physical rather than the social environment. I am borrowing, here, the language of the history and philosophy of science, a pivotal discipline in my view.

We have a horrific example of the danger of accepting that scientific truth about people has been established. Marxism claimed to be a scientific analysis of social history, and in its name Josef Stalin had millions killed. In Psychology, it is Gergen who has made very telling criticisms of any claim to provide a scientific analysis. This is the theme of my previous book

Psychology on the Couch, too. I concluded Chapter 1 of the present book with more points in this vein about the history of Psychology, the perspective from Clinical Psychology, philosophy of Psychology, and malfunctions of the publication process.

THE RISE OF ENVIRONMENTAL PSYCHOLOGY

Within Psychology, now a well-established subject, there has grown up a new branch called "Environmental Psychology". Though very many of its contributors are American, I suggest its rise is to be compared with the ascent of European "Green" movements, and of the idea of stewardship on a supra-national, perhaps planetary, scale. The effect of these movements has been to give a new meaning to our everyday understanding of the term "environment". This is no more than a modern twist to a very traditional viewpoint. This viewpoint is the idea that behaviour can be explained, with sufficient insight into its context; that we are *made* what we are. Obvious though it must seem to some, this idea has struggled through the centuries with the philosophy expressed by British Prime Minister John Major that we "should be less ready to understand, and more ready to condemn".

In this conservative philosophy, "deviants", of the various kinds, are *born* what they are, afflicted, if this school chooses to push its analysis any further, by congenital brain abnormalities.

The underlying agenda is that all effort spent on deviants should go into managing them. Rehabilitation is unrealistic. This saves the professional charged with managing deviance much labour. I wonder whether this saving of exertion does not supply much of the energy supporting this ideology.

Along with such ideas about deviants goes, very commonly, a notion that "their seed is cursed" and best destroyed, the "eugenic" programme.

I moved on in Chapter 2 to describe how this branch of academic study emerged. Some general features characterising Environmental Psychology are the "systems" approach, the applied orientation, and multi-disciplinarity. Classical psychologists who foreshadowed this branch include Brunswik, Lewin, Gibson, and Kohler & Wallach. The most prominent early areas of empirical research were the effects of architectural design, particularly of psychiatric wards; "proxemic" needs for personal space and territory; and abilities in forming images of one's town. Since then, the branch has expanded greatly. The political sensitivity of some issues studied has always had to be reckoned with.

I described some topic areas to give a flavour of the branch today. An area I did no more than touch upon is that of attempts to present a general theory about the psychological effects of the environment.

One substantive area that I considered in some detail, is environmental perception. There are some philosophical preliminaries to deal with. Then, I mentioned the work of von Senden on cataract removal, and of Fantz on neonate face recognition. I discussed the phenomenon of habituation, and Gregory's work on the dependence of visual illusions on a straight-line built environment. There is also the importance of speech context in phonetic perception.

Next, I went on to the effects of the earth's atmosphere. The wild speculations of Huntingdon about climate lead in to more empirically grounded ideas about the effects of heat, cold, wind and pollution.

My major attempt to give the flavour of the field, though, was a reconstruction of some contemporary research, in terms of the "town vs country" focus.

I described some modern research following on from the early "urban imagery" work, now often called "cognitive mapping", such as some very interesting work on "neighbourhoods", and on featureless environments like underground metros. This ties in with classical theorising (of Tolman) derived from animal Psychology.

Work on crowding also draws some of its inspiration from animal models, like the lemmings, and experimental "behaviour sinks". The Malthusian tradition, of worry about the environment's capacity to sustain population, continues today.

The phenomenon of animal territoriality, has also been generalised by psychologists to human behaviour. We speak of people's primary, secondary and public territories, but human territoriality is clearly utterly distinctive, in the importance for it of cultural factors. Individual factors are also consequential. It is highly aversive to have one's personal space "invaded". National cultures vary in what they regard as the desirable distance for interpersonal interaction.

THE DETRIMENTAL ENVIRONMENT: THE QUESTION OF EVALUATION

In this book, I have sought to relate Environmental Psychology to problems of mental health. When trying to explain mental problems, it is necessary to have some idea of what forms a "bad" or detrimental environment. The purpose of Chapter 3 was to explore *evaluation* of this sort. This has been the area of my own research within the field. Rather than try, and fail, to cover the huge area of evaluation comprehensively, I restricted myself in Chapter 3 to specialised topics. Thereby, I hoped to give an idea of how much further work could be done to investigate the many general issues I raise throughout the book.

I began with a little on "local identification", then said something to show how evaluation has depended on the historical context, as with the historical movement known as Romanticism.

Next, I described some general Social Psychology of *attitudes*, since these are basic to evaluation. I explained the theory of "cognitive dissonance" as the source of attitude change, as well as some ideas about how we form attitudes in the first place. I regard understanding how attitudes cluster together as a major project for future research.

In tune with the "green" tinge to Environmental Psychology, I presented a discussion of *ecological* attitudes, relating them back to the town vs country divide. There has been previous work on the measurement and analysis of such attitudes, showing that the general level of ecological awareness is painfully low. A study of my own found that rural children are more aware than urban ones, though it will make a difference whether one is looking at fertiliser pollution, or at littering and graffiti. There was also a gender difference, in that girls seem to care more about the environment but also take less action on the problems.

I then discussed the effect on its evaluation of the *familiarity* of an environment. This is an area where laboratory experimentation with different kinds of "stimuli" is possible. This has confirmed the common observation that environments "grow on you", what Psychology calls the "mere exposure effect". More recently, the experimental literature has focused also on *over-exposure* effects, where environments start to become stale and irksome. Much preliminary work has been done confirming that these effects depend greatly on the particular type of stimulus in question.

I have stressed that where the "environment" is other people (what I have been calling the social as opposed to the physical environment), questions of a wholly different order of complexity arise. Nevertheless, if it can be extended from the physical to the social, this sort of work has a bearing on the key idea that reconciliation of intergroup conflict will follow from more intergroup contact.

To build a bridge between the laboratory stimulus effects and the socio-political issue, I have done some work on one type of macro-scale physical environment, the natural landscape. This sort of investigation rides with the popularity of tourism, an increasingly important industry throughout the world. Even from the empirical research, one can see that liking or preference for a particular landscape is, of course, a far more complicated matter than simply assessing its familiarity. Nevertheless, using nature photographs, I have substantial evidence for effects of, first, growing attraction, and then, over-exposure, with increasing familiarity of the scene depicted. A brief discussion of some theories of landscape preference illustrates the complexity of the issues arising when appraising a real-life environment. It remains completely open, to my knowledge, whether the urban landscape can, perhaps in its cumulative impact, have a detrimental effect upon mental health.

ENVIRONMENTAL GENESIS OF MENTAL ILLNESS

Although using the term mental illness for identification, I agree with Szasz that it is a "category error", with terms like "mental disturbance" to be preferred. A similar point is made about the idea of "psychiatric diagnosis", with reference to the theory that these terms function merely as labels. This involves a discussion of the "antipsychiatric" critique by Laing and his associates of the "medical model", which focuses particularly on the physical treatments that are used in the name of that model. One idea of Laing that remains provocative today is that mental disturbance is quite natural. It is an adaptive response to adversity or other problems in living, a form of healing, that should be allowed to ride its course, and not be "short-circuited" by physical treatments, like drugs and electro-convulsive therapy.

The pivotal word in debating the medical model is "schizophrenia" or, even more, "a schizophrenic". Conventionally, doctors use it of people suffering a cognitive disturbance, such as delusion or hallucination. It is one of the most stigmatising terms imaginable. Some of the people referred to live and die in the most harrowing circumstances. Yet, when we take a recent book *Schizophrenia Genesis* (Gottesman, 1992), there is plenty of evidence, that the term is not used with much precision or objectivity. This has not prevented brain scientists from claiming that they have discovered the cause of this "disease".

It is the role of stress, in producing a disturbance, that needs to be appraised. It is understood that disturbance *can* be produced by a variety of organic pathologies, and conversely, that it *can* be an obvious reaction to stress. Where there is no clear cause, what sort of explanation do we prefer?

I feel two points need to be emphasised. Many practising psychiatrists take a causal role for stress for granted. And the latent potency of stress should be revealed to us by a phenomenon like brainwashing.

Research seeking causes in "life events" faces a formidable problem of complexity; it is very difficult to know what sort of stress is critical for a particular individual. Because disturbance can be responsible for negative life events, it is not always clear that the unemployment, say, produced the mental crisis, rather than vice versa. None the less, I should say that there is now a consensus that some form of loss often causes depression, in particular.

Also, there is now a substantial body of research, following up the victims of disasters of various kinds, and demonstrating that psychopathology occurs with much-above-baseline frequency. It seems to me, that the practical problem for psychiatrists, is whether they are successfully identifying all cases of "post-traumatic stress disorder" (an acknowledged diagnostic category).

I explore further the situation where a bereavement causes depression. I point out that mood depression commonly carries with it a certain type of cognitive disturbance, which I describe in a little detail. Bereavement reminds us of our own mortality, an insoluble problem that we all share. Separation from his mother in childhood, though temporary, may seem like a bereavement to the child. I consider the role of this in sensitising people to later loss (which is John Bowlby's "big idea"). Of course it is desirable to have a less stigmatising diagnosis like depression, but, in the end, can affective disturbance like this really be cleanly separated from cognitive disturbance like schizophrenia?

I consider, a little, the role of poverty in causing disturbance. There is certainly an *association* with diagnosed schizophrenia, but it has often been argued that this is because this disturbance itself reduces earning power. It is likely that poverty compels crowded, or otherwise unsatisfactory, housing conditions that cause mental disturbance. It is also likely that poverty goes with a loss of power to avoid a stigmatising diagnosis like "schizophrenia".

I further consider the importance of social support in rehabilitating someone from mental disturbance. This is relevant to the policy of "community care" for such people, and to the desirability or otherwise of returning them to their families. Such issues ride with the recognition that much of the disturbance is "iatrogenic" in nature, caused by the very institutions and the very medications intended to help.

Finally, I return to Environmental Psychology, to make a start on the question whether the city is a psychologically pathological environment. I include issues like rehousing, high-rise living (and the idea of non-functional factors in design choice), and environmental noise. I also refer to the idea of the "healing countryside", and to physical health aspects. It is impossible, really, to generalise, particularly when we compare cities in the First and Third Worlds. Moreover, the drawbacks of rural life also need some emphasis.

THE THERAPEUTIC ENVIRONMENT FOR THE MENTALLY UNHEALTHY

Care works, in practice, within fixed norms for the number of hospital beds available. The old mental institution, or "asylum", where, once, most of the care was provided, has come under increasing fire, though. Barton's

(1976) book on *Institutional Neurosis* has been taken very seriously in Britain, and a great deal done to try to cope with a disturbance whose core was apathy. In particular, there is now a policy of "community care", though (like most major innovations) it is very controversial. There remain General Hospital wards for people undergoing acute mental crises, and institutionalism is still a problem there. Too many mental health professionals are "hit-and-runners", who do not form a continuing relationship with the user. So they do not appreciate the importance to him of his physical environment, in spite of the growing body of empirical research confirming it. Thus, there is now a fair amount of research on the arrangements for an interior that ease good interpersonal interaction. There is a perennial question about staffing levels. I have also said a little about the way certain physical environments seem conducive to certain therapeutic orientations, with some context about how Psychotherapy relates to mainstream basic Psychology.

The very appearance and situation of the asylums hints at a custodial function, and, to some extent, they have been designed for that function to the detriment of therapy. This reflects the existence of mental health law, which, in my view, damages the user's relationship with staff, and is highly counter-therapeutic. It is also highly questionable, whether the law keeps up with therapeutic developments effectively.

I give some further examples of dysfunctionality of current design, and consider whether this *would* be an appropriate area for the courts to be involved in. As user apathy is such a central part of their problem, it would be wrong to expect every instance of poor design to be revealed by their complaints.

With regard to the design of General Hospital psychiatric units, a Government minister has acknowledged that, "we do not yet have a prototype for the future". There are critics of the very idea of such units.

Community care, though a partial answer to the problem of institutional psychology, is difficult to set up in practice. There is a need to take heed that *some* institutions are "centres of excellence", and should be protected. Evaluation in action of the policy of community care, which is still in its early days in Britain, gives mixed feedback. I have presented some experience of my own, and the results of some research of my own, emphasising how much depends on the character of the individual community. I also mention some research in France.

I have presented work on making architectural design more user-friendly. This may imply aesthetic qualities as well as functionality. I describe some work of Neil McDougall, which is ongoing, including the ideas of "privacy gradients", liberation from engineering constraints, and "confusion potential". I also discuss an idea that he shares with the Canters, that the asylum should be a new home for the user.

MENTAL RETARDATION—EXPLANATION AND CARE

With "mental retardation", too, there are problems of language, and I use the newer term, "learning disability". It is likely that there is some overlap in who this covers with "mental disturbance".

I have said a little about terms like "constitutional" factors, "minimal brain damage" and "intelligence". There is a good deal of potential for mystification here, which encourages the harmful idea that there is some "hidden key" to learning disability.

I have covered some problems in testing intelligence, taking the standpoint that it remains a valuable, and perhaps necessary, activity. I am, however, skeptical about the progress of work on the physical (cerebral) basis of intelligence. I give a brief account of the controversy over the work on intelligence by Sir Cyril Burt.

It is likely that peri-natal events play a part in later learning disability. Such events may be more common among the poor. This is one reason deprived upbringing may be a factor in explaining the disability. The idea of a self-feeding cycle, in which deficits have consequences that lead to further deficits, is important in environmentalist theory.

I have considered work on the genetics of intelligence, acknowledging that there is evidence for a genetic component, though some arguments are fallacious. Even taking the family studies at face value, the old figure of 80% is giving way to one of 60%. This idea of a figure for "heritability" is frequently misunderstood, as applying to individuals rather than to groups.

I give close attention to one effort to show a biological link to intelligence that I have tried myself, namely studies of the effects of handedness. Clinical neurology has long ascribed a special significance to this, and there have been many reports, among which Levy (1969) is noteworthy, of cognitive correlates. Such correlates do not necessarily reflect neurological differences, but rather the minority status of left-handers. At the same time, their minority status may have neurological as well as environmental corollaries.

The second half of Chapter 6 considers the care of the learning disabled. I consider more discussion is necessary about the provision of opportunities for sexual intercourse and parenthood in this group. Clearly, a geneticist orientation towards explanation of the disability will encourage a more restrictive stance. Public silence on this issue is one facet of a more general silence, about the care of the learning disabled, that seems to me to be the immediate current problem.

There is more than one reason why frustration should develop into intellectual conviction that education has limited scope with the learning disabled. One is the easy misunderstanding of the implications of the evidence on "heritability" of intelligence. Another is an overemphasis on the importance of early experience for later personality, though this is an intellectual divide that I find it difficult to resolve for myself, even.

There is good evidence from a variety of sources that improving the upbringing of children does have a substantial effect upon intelligence quotients, though the problems with these have to be borne in mind.

The problems with institutional care for those mentally disturbed apply *pari passu* to the learning disabled. More target-specific educational programmes may transform the learning climate.

Where such children have behaviour problems, on top of their learning disability, it may be necessary to be tough-minded (in terms of locks, partitions, etc.) about the arrangement of their physical environment, though the ultimate priority of user needs has to be remembered. In practice, some buildings, even for "community care", look like prisons, and deter interaction with the community. Integration within a neighbourhood probably dictates a home should be small, and, in turn, specialised. Environments should ease the learning of "social skills" (though this is a lifetime project for all of us). "Normalisation" is an important theme in modern thinking about learning disability, and part of that is resisting the temptation to be *too* helpful.

ARE WE BORN OR MADE?

The nature/nurture debate is entwined with political philosophy. Acceptance of innateness and destiny goes with fatalism, which I see as allied to conservatism. Conservatives, who may benefit more from legal inheritance, are also likely to pay more attention to the possibilities of biological inheritance.

To my mind, the most politically sensitive differences are those of class, race and gender. Class and snobbery are British obsessions, and need a book to themselves. Race relations have required legislation in Britain, and it is wrong to inflame them by exaggerating the genetic differences. Sex discrimination legislation has also been necessary, and the recent past has revealed amply that occupational barriers justified partly by nativist arguments were malign.

Tuberculosis was once thought to be hereditary. Slum conditions were alleviated, and it disappeared. An underclass is reappearing, and with it tuberculosis. Poverty can be reduced, and in consort the diseases of poverty.

These issues do not all go together. Though I have argued that there is a systematic bias in science towards nativism, it may yet be that in specific areas of contention the nativist position will turn out to be correct.

In the last resort, we turn for our answers on issues like nature/nurture to the academic, and particularly the eminent academic. Whether such eminence is itself achieved by means of innate qualities is more than just another aspect of the wider issue. If the academic explains her own productivity in nativist terms, she is likely to favour nativist accounts of the issues she is studying.

In my view, motivation, rather than ability alone, is basic to explaining productivity. I should favour explaining high motivation in terms of early parental interactions. I list many other factors in the personal background of scientists, uncovered by the work of Roe, particularly, and of Zuckerman. Simonton has done some parallel historical work on the effects of the sociopolitical context.

The most frightening manifestation of malign nativism has been Adolf Hitler's belief in hereditary taint. This took in many medical conditions and forms of deviance, as well as non-"Aryan" races such as the Jews and Slavs, with positive discrimination in favour of the pure Aryan race. Hitler's State sterilised many people under legal compulsion. Under conditions of war, such ideas led on to the "final solution" of the "Jewish problem" by means of genocide. As I write, such "ethnic cleansing" is reappearing in Europe.

In a broader and milder sense, "eugenic" ideas, on influencing others whether they have children, have a long history, given a particular stimulus by misapplication of Darwin's theory of biological evolution through natural selection. Today, there is a school, known as Social Biology, that attempts in similar ways to extend modern findings in zoology to human society. Even criminal behaviour has been interpreted genetically. There is an acknowledged place for genetic counselling for couples at risk of having a child with a single-gene Mendelian disorder. When mental and learning disorders are interpreted, contentiously, as "polygenic", such counselling becomes decidedly sinister.

The main evidence that behaviours are genetic comes from family or "kinship" studies. I offer a critique of the evidence. I argue that it does not take proper account of basic genetic theory. I suggest some alternative environmental interpretations of findings taken to show a genetic component. I criticise the quality of the data. The twin evidence presents particular logical and data-quality problems. So do adoption studies.

I regard much explanation of behaviour in terms of the brain as no more than "neuromythology". To regard such evidence as supporting the existence of genetic mechanisms is to lean on a crutch that isn't there.

CONCLUSION—THE ATTRIBUTION ERRORS

When I give a cause for an event, I am *attributing* the event to something. A plant dies in the garden, and I attribute it to lack of water. A man collapses in the street and you attribute it to a heart attack, or alternatively to the effects of alcohol. The process of attribution is a major focus of Social Psychology. Once, it accounted for more research papers than any other specific topic (Baron & Byrne, 1991; Eiser, 1980; Heider, 1958).

The first achievement of this research, and the one directly relevant to the theme of this book, was the identification of a "fundamental attribution error". We are all much too prone to see behaviour as caused "internally" rather than "externally": for example as due to personality rather than circumstances. This bias towards the internal applies much more to our explanations of others than to our explanations of ourselves. We recognise the influence of situations upon our own behaviour far more than we recognise it in those whom we observe. This is known as the "actor–observer difference in attribution". These are research terms for faults that most thinking people recognise anyway: that we apply different standards for ourselves, that we find it difficult to put ourselves in someone else's place.

These errors are enough to account for the mistaken popularity of hereditarianism. The view that other people are "just like that", genetically, is really just a familiar cognitive error.

Moreover, I should say, the harder we find it to relate personally to those suffering mental disturbance or learning disability, the readier we shall be to attribute their condition to internal causes. It is a variant of the actor–observer difference. Difficulties in relating in this way are partly due to lack of personal experience of the problems, but there is another explanation. It is so common to find that the member of a family who really needs mental help is the one who would "not touch a psychotherapist with a bargepole". In a way, the strength of the resistance is a sign of the depth of the underlying problems. Readiness to recognise our own needs is the beginning of understanding of others.

Once errors have been identified, they can be rooted out. We live in an age of steadily increasing prosperity and leisure time that can be, at least partly, time for reflection. Is it too optimistic to hope that people will find it easier and easier to empathise with their less fortunate neighbours? To me, this will be a sign of true progress.

REFERENCES

Baron RA & Byrne D (1991). *Social Psychology: Understanding Human Interaction, 6th edn.* London: Allyn & Bacon.

Eiser JR (1980). *Cognitive Social Psychology: A Guidebook to Theory and Research.* London: McGraw-Hill.

Heider F (1958). *The Psychology of Interpersonal Relations.* New York: John Wiley. Heider used to keep comprehensive workbooks about his own social encounters. What some people will do for learning!

Bibliography

Advances in Environment Psychology [Various editors]. Hillsdale, NJ: Erlbaum.
Albrecht D, Bultena G, Holberg E & Nowak P (1982). The New Environmental Paradigm: measuring environmental concern. *Journal of Environmental Education 13,* 39–43.
Ammons RB (1987). Psychology from the standpoint of a mechanist: the early life and work of Clark L Hull. Unpublished PhD dissertation, Duke University.
Annett M & Manning M (1989). The disadvantages of dextrality for intelligence. *British Journal of Psychology 80,* 213–226.
Annett M & Manning M (1990). Arithmetic and laterality. *Neuropsychologia 28,* 61–69.
Annett M & Turner A (1974). Laterality and the growth of intellectual abilities. *British Journal of Educational Psychology 44,* 37–46.
Appleton J (1975). *The Experience of Landscape.* London: John Wiley.
Arcury TA & Johnson TP (1987). Public environmental knowledge: a statewide survey. *Journal of Environmental Education 18,* 31–37.
Arieti S (1974) (Ed) *Interpretations of Schizophrenia I.* New York: Basic Books.
Aronson E & Mills J (1959).The effect of severity of initiation on liking for a group. *Journal of Abnormal & Social Psychology 12,* 16–27.
Arthur LM (1977). Predicting scenic beauty of forest environments: some empirical tests. *Forest Science 23,* 151–160.
Ayer AJ (1969). *Foundations of Empirical Knowledge.* London: Macmillan.
Bacon F (1620). *Novum Organum.*
Barham P (1992). *Closing the Asylum: the Mentally Ill in Society.* Harmondsworth: Penguin.
Barham P & Hayward R (1991). *From the Mental Patient to the Person.* London: Routledge.
Barker R (1968). *Ecological Psychology: Concepts and Methods for Studying the Environment of Human Behavior.* Stanford, CA: Stanford University Press.
Barnes B (1985). *About Science.* Oxford: Blackwell.
Baron RA & Byrne D (1991). *Social Psychology: Understanding Human Interaction, 6th edn.* London: Allyn & Bacon.
Barton R (1976). *Institutional Neurosis, 3rd edn.* Bristol: John Wright.
Bayes K (1967). *The Therapeutic Effect of Environment on Emotionally Disturbed and Mentally Subnormal Children.* Old Woking: Unwin.
Bean P (1980). *Compulsory Admissions to Mental Hospitals.* Chichester: John Wiley.
Bean P & Mounser P (1993). *Discharged from Mental Hospitals.* London: Macmillan.
Beck AT (1976). *Cognitive Therapy and the Emotional Disorders.* Harmondsworth: Penguin.

Ben-David J (1962). Productivity and academic organisation in nineteenth-century medicine. In B Barber & W Hirsch, *The Sociology of Science*. New York: Free Press.
Ben-David J & Sullivan TA (1975). Sociology of science. *Annual Review of Sociology 32,* 203–222.
Benton AL (1964/5). Contributions to aphasia before Broca. *Cortex 1,* 314–327.
Berman M (1984). *The Re-enchantment of the World.* New York: Bantam Books.
Berlyne DE (1971). *Aesthetics and Psychobiology.* New York: Appleton–Century–Crofts.
Berquier A & Ashton R (1991). A selective review of possible neurological etiologies of schizophrenia. *Clinical Psychology Review 11,* 645–661.
Binet A & Simon T (1908). The development of intelligence in the child. *Année Psychologique 14,* 1–90.
Blakemore C, Iversen SD and Zangwill OL (1972). Brain functions (asymmetry of cerebral hemisphere function). *Annual Review of Psychology 23,* 433–456.
Bloom BS (1975). Mastery learning and its implications for curriculum development. pp 334–350 In M Golby, J Greenward & R West *Curriculum Development.* London: Croom Helm.
Blum A (1987). Students' knowledge and beliefs concerning environmental issues in four countries. *Journal of Environmental Education 18,* 7–13.
Bossard JH (1931). Residential propinquity as a factor in marriage selection. *American Journal of Sociology 38,* 219–224.
Bowlby J (1973). *Attachment and Loss. Vol. 2. Separation.* New York: Basic Books.
Boyers R & Orrill R (1972). *Laing and Antipsychiatry.* Harmondsworth: Penguin
Bradshaw JL & Nettleton NC (1983). *Human cerebral asymmetry.* New York: Appleton–Century–Croft.
Bradshaw JL, Nettleton NC, Taylor MJ & Mercelito J (1981). Right hemisphere language and cognitive deficit in sinistrals? *Neuropsychologia 19,* 113–132.
Brentano, Franz (1874). *Psychology from an Empirical Standpoint.* 1981 edition, London: Routledge & Kegan Paul.
Bridgman PW (1959). *The Way Things Are.* Oxford: Oxford University Press.
Breslau N, Davis GC, Andreski P & Peterson E (1991). Traumatic events and post-traumatic stress disorder in an urban population of young adults. *Archives of General Psychiatry 48,* 216–222.
Broca P (1861). Remarques sur le siège de la faculté du langage articulé, suivies d'une observation d'aphème. *Bulletin de la Société Anatomique de Paris 6,* 398–407.
Brown JAC (1963). *Techniques of Persuasion.* Harmondsworth: Penguin.
Brown GW & Harris T (1978). *Social Origins of Depression.* London: Tavistock.
Bruner J (1983). *In Search of Mind.* New York: Harper & Row.
Bryden MP (1986). Dichotic listening performance, cognitive ability and cerebral organisation. *Canadian Journal of Psychology 40,* 445–456.
Bryden MP & Rainey CA (1963). Left–right differences in tachistoscopic recognition. *Journal of Experimental Psychology 66,* 568–571.
Buber M (1970). *I and Thou.* (trans. W Kaufmann, orig. work 1929) Edinburgh: T & T Clark.
Buffery AWH & Gray JA (1972). Sex differences in the development of spatial and linguistic skills. In C Ounsted & DC Taylor, *Gender Differences: their Ontogeny and Significance.* Edinburgh: Churchill Livingstone.
Buhyoff GJ & Riesenman MF (1979). Manipulation of dimensionality in landscape preference judgments: a quantitative validation. *Leisure Science 2,* 221–238.
Bull R (1975). The psychology of clothes and fashion. *Bulletin of the British Psychological Society 28,* 459–465.

Bullock A (1991). *Hitler and Stalin: Parallel Lives.* London: HarperCollins.
Burnett AA, Lane DM & Dratt LM (1982). Spatial ability and handedness. *Intelligence 6,* 57–68.
Burt C (1950). *The Backward Child, 3rd edn.* London: University of London Press.
Busfield J (1986). *Managing Madness.* London: Unwin Hyman.
Calhoun JB (1962). Population density and social pathology. *Scientific American 206,* 139–148.
Canter D & Canter S (1979a). Building for therapy. In D Canter & S Canter, *Designing for Therapeutic Environments: A Review of Research.* Chichester: John Wiley.
Canter D & Canter S (1979b). Creating therapeutic environments. In D Canter & S Canter, *Designing for Therapeutic Environments: A Review of Research.* Chichester: John Wiley.
Canter D & Canter S (1979c) *Designing for Therapeutic Environments: A Review of Research.* Chichester: John Wiley.
Cantor GN (1968). Children's "like–dislike" ratings of familiarized and unfamiliarized visual stimuli. *Journal of Experimental Child Psychology 6,* 651–657.
Charman DK (1980). Note on a failure to find hemispheric asymmetry for a small sample of strongly left-handed and right-handed males and females using verbal and visuospatial recall. *Perceptual & Motor Skills 51,* 139–145.
Charney E & Weissman MM (1988). Epidemiology of depressive illness. In JJ Mann, *Phenomenology of depressive illness.* New York: Human Sciences Press.
Chomsky N (1965). *Aspects of the Theory of Syntax.* Cambridge, MA: MIT Press.
Chomsky N (1992). *Chronicles of Dissent.* New York: AK Press.
Clare A (1976). *Psychiatry in Dissent.* London: Tavistock.
Claridge GS (1985). *Origins of Mental Illness: Temperament, Deviance and Disorder.* Oxford: Blackwell.
Clarke AM, Clarke ADB & Berg JM (1985) (Eds) *Mental Deficiency: The Changing Outlook, 4th edn.* London: Macmillan.
Cochrane R (1983). *The Social Creation of Mental Illness.* London: Longman.
Cohen D (1979). *JB Watson: The Founder of Behaviorism.* London: Routledge & Kegan Paul.
Cohen IB (1985). *Revolution in science.* Harvard University Press.
Colman AM, Sluckin W & Hargreaves DJ (1981). The effect of familiarity on preferences for surnames. *British Journal of Psychology 72,* 363–369.
Cone JD & Hayes SC (1980). *Environmental Problems/Behavioral Solutions.* Monterey, CA: Brook Cole.
Connolly C (1988). *Enemies of Promise.* London: Deutsch.
Connolly J (1985). Life happenings and illness. In RN Gaind, FI Fawzy, BL Hudson & RO Pasnau, *Current Themes in Psychiatry, Vol 4.* London: Macmillan.
Corballis MC (1983). *Human Laterality.* New York: Academic Press.
Coren S (1990). *Left-Handedness: Behavioral Implications and Anomalies.* Amsterdam: North Holland.
Coren S (1992). *The Left-Hander Syndrome: The Cause and Consequence of Left-Handedness.* New York: Free Press.
Crawford JR & Parker DM (1989). *Developments in Clinical and Experimental Neuropsychology.* New York: Plenum Press.
Crick F (1990). *What Mad Pursuit: A Personal View of Scientific Discovery.* Harmondsworth: Penguin.
Daniel TC & Boster RS (1976). Measuring Landscape Aesthetics: The Scenic Beauty Estimation Method. *USDA Forestry Service Research Paper RM-167.* Fort Collins, CO: Rocky Mountain Forest and Range Experimental Station.

Dearden P (1980). Landscape assessment: the last decade. *Canadian Geographer 24,* 316–325.
De Long MR & Saluoso-Deonier C (1983). Effect of redundancy on female observers' visual responses to clothing. *Perceptual & Motor Skills 57,* 243–246.
Diamond J (1991). *The Rise and Fall of the Third Chimpanzee.* London: Radius.
Dobzhansky T (1962). *Mankind Evolving.* London: Yale University Press.
Dubos R (1965) Science and Man's Nature. *Daedalus 94,* 223–244.
Dubos R (1972). *A God Within.* London: Scribner.
Ebbinghaus H (1908). *Abriss der Psychologie.* Leipzig: Veit.
Eisenson J (1962). Language and intellectual findings associated with right cerebral damage. *Language & Speech 5,* 49–53.
Eiser JR (1980). *Cognitive Social Psychology: A Guidebook to Theory and Research.* London: McGraw-Hill.
Elbert T & Birbaumer N (1987). *Individual Differences in Hemispheric Specialisation.* New York: Plenum Press.
Ellis A (1987). The impossibility of achieving consistently good mental health. *American Psychologist 47,* 364–375.
Ellis R & Whittington D (1986). *A Guide to Social Skills Training.* London: Croom Helm.
Ericksen MK & Sirgy MJ (1989). Achievement motivation and clothing behaviour: a self-image congruence analysis. *Journal of Social Behaviour & Personality 4,* 307–326.
Eysenck HJ (1988). Theories of personality. In AW Staats & LP Mos, *Annals of Theoretical Psychology 5.* New York: Plenum.
Eysenck HJ & Kamin LJ (1981). *Intelligence: The Battle for the Mind.* London: Macmillan.
Fairweather H (1982). Sex differences. In JG Beaumont, *Divided Visual Field Studies of Cerebral Organisation.* London: Academic Press.
Fennell E, Satz P, van den Abell T, Bowers D & Thomas R (1978). Visuospatial competency, handedness and cerebral dominance. *Brain & Language 5,* 206–214.
Festinger L & Carlsmith JM (1959). Cognitive consequences of forced compliance. *Journal of Abnormal & Social Psychology 58,* 203–210.
Feyerabend P (1975). *Against Method.* New York: Free Press.
Fisher JD, Bell PA & Baum AS (1984). *Environmental Psychology, 2nd edn.* Eastbourne: WB Saunders.
Fletcher R (1991). *Science, Ideology and the Media: The Cyril Burt Scandal.* New Brunswick, NJ: Transaction.
Forrest D (1974). *Francis Galton: The life and work of Victorian genius.* London: Paul Elek.
Foster HD (1992). *Health, Disease and the Environment.* London: Belhaven.
Fox MF (1983). Publication productivity among scientists: a critical review. *Social Studies of Science 13,* 285–305.
Freedman J (1975). *Crowding and Behavior.* San Francisco: Freeman.
Freeman H (1984). *Mental Health and the Environment.* London: Churchill Livingstone.
Freeman H (1989). Mental health and the urban environment. In R Krieps, *Environment and Health: A Holistic Approach.* Aldershot: Gower.
Fried M (1963). Grieving for a lost home. In LJ Duhl, *The Urban Condition: People and Policy in the Metropolis.* New York: Basic Books.
Frisch K von (1966). *Dancing Bees: An Account of the Life and Senses of the Honey Bee.* London: Methuen.

Ganzini L, McFarland BH & Cutler D (1990). Prevalence of mental disorders after catastrophic financial loss. *Journal of Nervous & Mental Disease 178,* 680–685.

Gardiner H (1985). *Frames of Mind.* London: Paladin.

Gergen KJ (1973). Social psychology as history. *Journal of Personality & Social Psychology 26,* 309–320.

Gibson JJ (1979). *An Ecological Approach to Visual Perception.* Boston: Houghton Mifflin.

Gide A (1925). *La Symphonie Pastorale.* Paris: Gallimard.

Gifford R, Hay R & Boros K (1982). Individual differences in environmental attitudes. *Journal of Environmental Education 14,* 19–23.

Giggs JA (1984). Residential mobility and mental health, In H Freeman, *Mental Health and the Environment.* London: Churchill Livingstone.

Gilgen AR (1982). *American Psychology since World War II: A Profile of the Discipline.* Westport, CT: Greenwood Press.

Gillie O (1976). *Who Do You Think You Are?* London: Hart-Davis.

Giroud, Françoise (1986). *Marie Curie: A Life,* trans. by L Davis. London: Holmes & Meier.

Gobineau JA, Comte de (1853). *Essay on the Inequality of Human Races.*

Goffman E (1968). *Asylums: Essays in the Social Situation of Mental Patients and Other Inmates.* Harmondsworth: Pelican.

Goldberg EM & Morrison SL (1969). Schizophrenia and social class. *British Journal of Psychiatry 109,* 785–802.

Goldberg W & Huxley P (1992). *Common Mental Disorders. A Bio-social Model.* London: Tavistock/Routledge.

Goldstein M & Goldstein I (1984). *The Experience of Science: An Interdisciplinary Approach.* New York: Plenum.

Gordon H (1923). Mental and scholastic tests among retarded children. *Board of Education Pamphlet No 44.* London: HMSO.

Gottesman II (1992). *Schizophrenia Genesis.* San Francisco: WH Freeman

Gould P & White R (1974). *Mental Maps.* Harmondsworth: Pelican.

Gould SJ (1985). *The Flamingo's Smile: Reflections in Natural History.* London: WW Norton.

Gove WR, Hughes M & Galle OR (1979). Overcrowding in the home: an empirical investigation of its possible pathological consequences. *American Sociological Review 44,* 59–80.

Gray JA (1985). The neuropsychology of anxiety. *Issues in Mental Health Nursing 7,* 201–228.

Gregory R (1966). *Eye and Brain: the Psychology of Seeing.* London: Weidenfeld & Nicolson.

Gunzberg HC & Gunzberg AL (1979). "Normal" environment with a plus for the mentally retarded. In D Canter & S Canter, *Designing for Therapeutic Environments: A Review of Research.* Chichester: John Wiley.

Hall ET (1959). *The Silent Language.* New York: Doubleday.

Hammitt WE (1983). The familiarity–preference component of on-site recreational experiences. *Leisure Science 4,* 177–193.

Hardyck C & Petrinovich LF (1977). Left-handedness. *Psychological Bulletin 84,* 385–404.

Hardyck C, Petrinovich LF & Goldman RD (1976). Left-handedness and cognitive deficit. *Cortex 12,* 266–279.

Hargie O, Saunders C & Dickson D (1987). *Social Skills in Interpersonal Communication.* London: Croom Helm.

Hargreaves DJ, Colman AM & Sluckin W (1983). The attractiveness of names. *Human Relations 36,* 393–401.
Harlow HF (1971) *Learning to Love.* San Francisco: Albion.
Harrison AA (1977). Mere exposure. In L Berkowitz, *Advances in Experimental Social Psychology.* New York: Academic Press.
Hearnshaw LS (1979). *Cyril Burt, Psychologist.* London: Hodder & Stoughton.
Heather N (1976). *Radical Perspectives in Psychology.* London: Methuen.
Hebb DO (1987). *A Textbook of Psychology, 4th edn.* by DC Donderi. Hove: Lawrence Erlbaum.
Heber R (1962). The concept of mental retardation: definition and classification. *Proceedings of the London Conference on Scientific Studies of Mental Deficiency 1,* 236–242.
Hécaen H & Piercy M (1956). Paroxysmal dysphasia and the problem of cerebral dominance. *Journal of Neurology, Neurosurgery & Psychiatry 19,* 194–201.
Heider F (1958). *The Psychology of Interpersonal Relations.* New York: Wiley.
Heilbron JL (1986). *Dilemmas of an Upright Man: Max Planck as Spokesman for German Science.* San Francisco: University of California Press.
Henle M (1986). *1879 and All That: Essays in the Theory and History of Psychology.* New York: Columbia University Press.
Herbst PG (1970). *Behavioural Worlds: the Study of Single Cases.* London: Tavistock.
Hermann DJ & Van Dyke KA (1978). Handedness and the mental rotation of perceived patterns. *Cortex 14,* 521–529.
Herrnstein RJ (1973). *IQ in the Meritocracy.* London: Allen Lane.
Herzog TR, Kaplan S & Kaplan R (1976). The prediction of preference for familiar urban places. *Environment & Behavior 8,* 627–645.
Hicks RA & Beveridge R (1978). Handedness and intelligence. *Cortex 14,* 304–307.
Hicks RA & Dusek CM (1980). Handedness distributions of gifted and non-gifted children. *Cortex 16,* 479–481.
Hicks RA & Kinsbourne M (1978). Human handedness. In M Kinsbourne, *Asymmetrical Function of the Brain.* Cambridge: Cambridge University Press.
Holahan CJ (1979). Environmental psychology in psychiatric hospital settings. In D Canter & S Canter, *Designing for Therapeutic Environments: A Review of Research.* Chichester: John Wiley.
Holmes TH & Rahe RH (1967). The Social Readjustment Rating Scale. *Journal of Psychosomatic Research 11,* 213–218.
Honnold JA (1984). Age and environmental concern: some specification of effects. *Journal of Environmental Education 16,* 4–9.
Hull RB & Buhyoff GJ (1983). Distance and scenic beauty: a nonmonotonic relationship. *Environment & Behavior 15,* 77–92.
Hummell CF (1977). The effects of induced cognitive sets in viewing air pollution scenes. Unpublished doctoral dissertation, Colorado State University.
Hunt J McV (1961). *Intelligence and Experience.* New York: Ronald Press.
Illich I (1971). *Deschooling Society.* New York: M Boyers.
Illich I (1991). *Limits to Medicine.* Harmondsworth: Penguin.
Ineichen B (1979). High rise living and mental stress. *Biology & Human Affairs 44,* 81–85.
Inomata S (1983). Research on repeated exposure effects of the stimulus. *Japanese Journal of Experimental Social Psychology 23,* 39–52.
Ittelson WH, Proshansky HM, Rivlin LG & Winkel GH (1974). *An Introduction to Environmental Psychology.* New York: Holt, Rinehart & Winston.

Jackson RH, Hudman LE & England JL (1978). Assessment of the environmental impact of high voltage power transmission lines. *Journal of Environmental Management 6,* 153–170.

Jaki SL (1978). *The Origin of Science and the Science of its Origin.* Edinburgh: Scottish Academic Press.

Jariabkhova K (1980). Pokazateli intellektualhogo razvitiya u pravoruchnykh i levoruchnykh dety [Intellectual achievements by right- and left-handed children]. *Studia Psychologica 22,* 249–253.

Jaus HH (1984). The development and retention of environmental attitudes in elementary school children. *Journal of Environmental Education 15,* 33–36.

Jeffcoate WJ (1985). Enkephalins, endorphins and psychiatric disease. In RN Gaind, FI Fawzy, BL Hudson & RO Pasnau, *Current Themes in Psychiatry, Vol 4.* London: Macmillan.

Jencks C (1973). *Inequality.* Harmondsworth: Penguin.

Jensen AR (1969). How much can we boost IQ and scholastic achievement? *Harvard Educational Review 39,* 1–123.

Jodelet D (1991). *Madness and Social Representations.* Hemel Hempstead: Harvester Wheatsheaf.

Johnson O & Harley C (1980). Handedness and sex differences in cognitive tests of brain laterality. *Cortex 16,* 73–82.

Johnson P (1993). *Intellectuals.* London: Orion.

Johnstone L (1989). *Users and Abusers of Psychiatry.* London: Routledge.

Jones MC (1983). *Behaviour Problems in Handicapped Children: The Beech Tree House Approach.* London: Souvenir.

Jorgensen BW & Cervone JS (1978). Affect enhancement in the pseudorecognition paradigm. *Personality & Social Psychology Bulletin 4,* 285–288.

Joynson RB (1989). *The Burt Affair.* London: Routledge.

Kallman FJ (1938). *The Genetics of Schizophrenia.* New York: Augustin.

Kamin LJ (1974). *The Science and Politics of IQ.* Potomac, MD: Erlbaum.

Kaplan R (1975). Some methods and strategies in the prediction of preference. In EH Zube, J Brush & R Fabos, *Landscape Assessment.* Stroudsburg PA: Dowden Hutchinson & Ross, pp118–129.

Kaplan S (1975). An informal model for the prediction of preference. In EH Zube, J Brush & R Fabos, *Landscape Assessment.* Stroudsburg PA: Dowden Hutchinson & Ross, pp92–107.

Kaplan S & Kaplan R (1982). *Cognition and Environment.* New York: Praeger.

Kaplan S, Kaplan R & Wendt JS (1972). Rated preference and complexity for natural and urban visual material. *Perception & Psychophysics 12,* 334–356.

Katz P (1937). *Animals and Men.* New York: Longman, Green.

Kavanagh JF (1988). *Understanding Mental Retardation: Research Accomplishments and New Frontiers.* Baltimore, MD: Paul Brookes.

Kellett JM (1984). Crowding and territoriality: a psychiatric view. In H Freeman, *Mental Health and the Environment.* London: Churchill Livingstone.

Kessler RC, Price, RH & Wortman CB (1985). Social factors in psychopathology: stress, social support and coping processes. *Annual Review of Psychology 36,* 531–572.

Kety SS, Rosenthal D, Wender PH, Schulsinger F & Jacobsen B (1978). The biological and adoptive families of adopted individuals who become schizophrenic. In LC Wynne, RL Cromwell & S Matthysse, *The Nature of Schizophrenia.* New York: John Wiley.

Kimura D (1966). Dual functional asymmetry of the brain in visual perception. *Neuropsychologia 4*, 275–285.
King J (1987). A review of bibliometric and other scientific indicators and their role in research evaluation. *Journal of Information Science 13*, 261–76.
Kitterle FL (1991). *Cerebral Laterality: Theory and Research: The Toledo Symposium*. New York: Lawrence Erlbaum.
Kocel KM (1977). Cognitive abilities, handedness, familial sinistrality and sex. *Annals of the New York Academy of Science 299*, 233–241.
Kohler W & Wallach H (1944). Figural after-effects: an investigation of visual processes. *Proceedings of the American Philosophical Society 88*, 269–357.
Kovac D, Jariabkova K & Zapotocna O (1984). Semilongitudinal study of laterality, cognition and personality. *Studia Psychologica 26*, 71–74.
Kraepelin E (1887). *Psychiatrie, 2nd edn*. Leipzig: Abel.
Kreimer A (1977). Environmental preference: a critical analysis of some research methodologies. *Journal of Leisure Research 9*, 88–97.
Kuhn TS (1962). *The Structure of Scientific Revolutions*. Chicago University Press.
Kuipers B (1982). The "Map in the head" metaphor. *Environment & Behavior 14*, 202–220.
Kuipers L (1992). Expressed Emotion research in Europe. *British Journal of Clinical Psychology 31(4)*, 429–443.
Kupfersmid J (1988). Improving what is published: a model in search of an editor. *American Psychologist 43*, 635–642.
Laing RD (1967). *The Politics of Experience*. Harmondsworth: Penguin.
Laing RD & Esterson A (1965). *Sanity, Madness and the Family. Vol 1. Families of Schizophrenics*. New York: Basic Books.
Lakatos I & Musgrave AE (1970). *Criticism and the Growth of Scientific Knowledge*. Cambridge: Cambridge University Press.
Lamb R (1979). *Alternatives to Acute Hospitalisation*. San Francisco: Jossey-Bass.
Leahey TH (1987). *A History of Psychology, 2nd edn*. Englewood Cliffs, NJ: Prentice-Hall.
Lee T (1976). *Psychology and Environment*. London: Methuen.
Lee TR (1954). "Neighbourhood" as a socio-spatial schema. Unpublished doctoral dissertation of the University of Cambridge.
Leff J (1983) *Planning a Community Psychiatric Service: From Theory to Practice*. London: Royal College of Psychiatrists
Leighton DC, Harding JS, Macklin DB, Hughes CC & Leighton AH (1963). Psychiatric findings of the Stirling County Study. *American Journal of Psychiatry 119*, 1021–1026.
Lévi-Strauss C (1992). *Tristes Tropiques*. Harmondsworth: Viking Penguin.
Levy J (1969). Possible basis for the evolution of lateral specialisation of the human brain. *Nature 224*, 614–615.
Levy-Leboyer C (1982). *Psychology and Environment*. London: Sage
Lewicki P (1982). Social psychology as viewed by its practitioners: survey of SESP members' opinions. *Personality & Social Psychology Bulletin 8*, 409–416.
Lewin K (1951). *Field Theory in Social Science: Selected Theoretical Papers*. New York: Harper & Brothers.
Lindeboom GA (1975). *Letters of Jan Swammerdam to Melchisedec Thevenot*. New York: Swets.
Littlewood J (1992). *Aspects of Grief*. London: Routledge.
Livi-Bacci M (1992). *A Concise History of World Population*. Oxford: Blackwell.

Lombroso C (1895). *L'homme criminel, 2nd edn.* Paris: Ancienne Librairie Germer Baillihre et Cie.
Lorenz K (1966). *On Aggression.* London: Methuen.
Lorenz K (1981). *The Foundation of Ethology.* New York: Springer-Verlag.
Lynch K (1960). *The Image of the City.* Cambridge, MA: MIT Press.
Lyons E (1983). Demographic correlates of landscape preference. *Environment & Behavior 15,* 487–511.
Macaulay TB (1843). *History of England from the Accession of James II.* 1979 edition, (Ed) H Trevor-Roper, Harmondsworth: Penguin.
Machiavelli N (1908). *The Prince.* London: JM Dent (orig. work 1532).
Mackworth J (1970). *Vigilance and Attention: A Signal Detection Approach.* Harmondsworth: Penguin.
Mahoney MJ (1979). Psychology of the scientist: an evaluative review. *Social Studies of Science 9,* 349–375.
Maloney MP & Ward MO (1973). Ecology: let's hear from the people. An objective scale for the measurement of ecological attitudes and knowledge. *American Psychologist 28,* 583–585.
Maloney MP, Ward MO & Braucht CN (1975). A revised scale for the measurement of ecological attitudes and knowledge. *American Psychologist 30,* 787–790.
Mangen SP (1982). *Sociology and Mental Health: An Introduction for Nurses and Other Care-Givers.* London: Churchill Livingstone.
Maslow AH (1954). *Motivation and Personality.* New York: Harper & Row.
Masson J (1989). *Against Therapy.* London: Collins.
McDougall N (in prep.) The environment and mental health: the relevance of Chris Alexander's "pattern language". *Radical Health Promotion.*
McGlone J & Davidson W (1973). The relation between cerebral speech laterality and spatial ability with special reference to sex and hand preference. *Neuropsychologia 11,* 105–113.
McGlone J & Kertesz A (1973). Sex differences in cerebral processing of visuospatial tasks. *Cortex 9,* 313–320.
McKeever WF (1986). The effects of handedness, sex and androgyny on language laterality, and verbal and spatial ability. *Cortex 22* (4).
McNemar Q (1955). *Psychological Statistics, 2nd edn.* New York: Wiley.
McWhirter L (1983). Contact and conflict: the question of integrated education. *Irish Journal of Psychology 6,* 13–27.
Medawar P & Medawar J (1977). *The Life Sciences.* New York: Harper & Row.
Merton RK (1965). *On the Shoulders of Giants: A Shandean postscript.* New York: Free Press.
Miller E (1982). Handedness and a test of cognitive development. *Neuropsychologia 20,* 155–162.
Milner B (1972). Interhemispheric differences and psychological processes. *British Medical Bulletin 27,* 272–278.
Milner B, Branch C & Rasmussen T (1966). Evidence for bilateral speech representation in some non-right-handers. *Transactions of the American Neurological Association 80,* 42–57.
Milstein V, Small IF, Malloy FW & Small JG (1979). Influence of sex and handedness on hemispheric functioning. *Cortex 15,* 439–449.
Monroe SM & Simons AD (1991). Diathesis–stress theories in the context of life stress research: implications for the depressive disorders. *Psychological Bulletin 110,* 406–425.

Mulkay MJ (1979). *Science and the Sociology of Knowledge*. London: George Allen & Unwin.
Munsinger H (1975). The adopted child's IQ: a critical review. *Psychological Bulletin 82*, 623–659.
Needham J (1954–). *Science and Civilization in China*. 7 vols. Cambridge: Cambridge University Press.
Neisser U (1967). *Cognitive Psychology*. New York: Appleton–Century–Crofts.
Neisser U (1976). *Cognition and Reality: Principles and Implications of Cognitive Psychology*. San Francisco, CA: WH Freeman.
NHS Health Advisory Service and DoH Social Services Inspectorate (1991) *Report on Services for Mentally Ill People and Elderly People in the Torbay Health District*.
Olsen MR (1985). The care of the chronically mentally ill: boarding out an alternative to family and hospital care. In RN Gaind, FI Fawzy, BL Hudson & RO Pasnau, *Current Themes in Psychiatry, Vol 4*. London: Macmillan.
Opler LA, Ramirez PM, Rosenkilde CE & Fiszbein A (1991). Neurocognitive features of chronic schizophrenic inpatients. *Journal of Nervous & Mental Disease 179*, 638–640.
Orem DE (1971). *Nursing: Concepts of Practice*. New York: McGraw-Hill.
O'Riordan T (1976). Attitudes, behavior and environmental policy issues. In I Altman & JF Wohlwill, *Human Behavior and Environment: Advances in Theory and Research, Vol 1*. New York: Plenum.
Ortega y Gasset J (1929). *The Revolt of the Masses*. 1964 edition, London: WW Norton.
Osmond H (1957). Function as the basis of psychiatric ward design. *Mental Health* (Architectural Supplement) *8*, 23–29.
Owen D (1985). *None of the Above: Behind the Myth of Scholastic Aptitude*. Boston: Houghton Mifflin.
Pasewark RA, Fitzgerald BJ & Sawyer RN (1975). Psychology of the scientist: XXXII. God at the synapse: research activities of clinical, experimental and physiological psychologists. *Psychological Reports 36*, 671–674.
Pastore N (1949). *The Nature–Nurture Controversy*. New York: King's Crown Press.
Patsfall MR, Feimer NR, Buhyoff GJ & Wellman JD (1984). The prediction of scenic beauty from landscape content and composition. *Journal of Environmental Psychology 4*, 7–26.
Pearce P (1982). *Social Psychology of Tourist Behaviour*. Oxford: Pergamon.
Pearce SR & Waters NM (1983). Quantitative methods for investigating the variables that underlie preference for landscape scenes. *Canadian Geographer 27*, 328–344.
Peplau HE (1988). *Interpersonal Relations in Nursing*. London: Macmillan.
Peters DP & Ceci SJ (1982). Peer-review practices of psychology journals: the fate of published articles, submitted again. *Brain & Behavioral Sciences 5*, 187–255.
Peters M (1991). Reanalysis of Benbow's data on mathematical giftedness. *Canadian Journal of Psychology 45*, 415–419.
Popper KR (1963). *Conjectures and Refutations*. London: Routledge & Kegan Paul.
Porac C, Coren S & Duncan P (1980). Lateral preference in retardates: relationships between hand, eye, foot and ear preference. *International Journal of Clinical Neuropsychology 2*, 173–187.
Proshansky HM, Ittelson WH & Rivlin LG (1978). *Environment Psychology: Man and his Physical Setting*. New York: Holt, Rinehart & Winston.
Ramsey CE & Rickson RE (1976). Environmental knowledge and attitudes. *Journal of Environmental Education 8*, 10–18.

Read H (1949). *The Meaning of Art.* Harmondsworth: Pelican.
Reitan RM (1955). Certain differential effects of left and right cerebral lesions in human adults. *Journal of Comparative & Physiological Psychology 48,* 474–477.
Reitan RM (1966). Problems and prospects in studying the psychological correlates of brain lesions. *Cortex 2,* 127–154.
Renan E (1863). *Life of Jesus.* 1935 edition, London: Watts.
Richards S (1983). *Philosophy and Sociology of Science: an Introduction.* Oxford: Blackwell.
Roberts JM (1985). *The Triumph of the West.* London: Guild Publishing.
Roe A (1953). *The Making of a Scientist.* Westport, CT: Greenwood Press.
Rogers CR (1951). *Client-centred Therapy.* Boston: Houghton Mifflin.
Romney DM (1990). Thought disorder in the relatives of schizophrenics: a meta-analytic review of selected published studies. *Journal of Nervous & Mental Disease 178,* 481–486.
Ronco P (1972). Human factors applied to hospital patient care. *Human Factors 14,* 461–470.
Rose S, Lewontin RC & Kamin L (1990). *Not in our Genes: Biology, Ideology and Human Nature.* Harmondsworth: Penguin.
Rosenhan D (1975). On being sane in insane places. In T Scheff, *Labelling Madness.* Englewood Cliffs, NJ: Prentice-Hall.
Ruse M (1979). *Sociobiology: Sense or Nonsense?* Dordrecht: Reidel.
Ruse M (1989). *The Darwinian Paradigm: Essays on its History, Philosophy, and Religious Implications.* London: Routledge.
Russell B (1946). *History of Western Philosophy.* London: George Allen & Unwin.
Russell CA (1985). *Cross-currents: Interactions between Science and Faith.* Leicester: Inter-varsity Press.
Sanders B, Wilson JR & Vandenburg SG (1982). Handedness and spatial intelligence. *Cortex 18,* 79–80.
Satz P (1973). Left-handedness and early brain insult: an explanation. *Neuropsychologia 11,* 115–117.
Satz P & Fletcher JM (1987). Left-handedness and dyslexia: an old myth revisited. *Journal of Pediatric Psychology 12,* 291–298.
Schlenker BR (1974). Social psychology and science. *Journal of Personality & Social Psychology, 29,* 1–15.
Schultz DP & Schultz SE (1987). *A History of Modern Psychology, 4th edn.* London: Harcourt, Brace, Jovanovitch.
Seamon JG, Brody N & Kauff DM (1983). Affective discrimination of stimuli that are not recognised. II. Effect of delay between study and test. *Bulletin of the Psychonomic Society 21,* 187–189.
Searle JR (1992). *The Rediscovery of Mind.* Cambridge, MA: MIT Press.
Searleman A (1977). A review of right hemisphere linguistic capabilities. *Psychological Bulletin 84(3),* 503–528.
Sedgwick P (1982). *Psychopolitics.* London: Pluto Press.
Selye H (1978). *Stress of Life.* New York: McGraw-Hill.
Shafer EL Jr & Meitz J (1970). It Seems Possible to Quantify Scenic Beauty in Photographs. *USDA Forestry Service Research Papers NE-162.* Upper Darby, PA: North-East Forest Experimental Station.
Shafer EL Jr & Richards TA (1974). A Comparison of Viewer Reactions to Outdoor Scenes and Photographs of those Scenes. *USDA Forestry Service Research Papers NE-202.* Amherst, MA: North-East Forest Experimental Station.

Shafer EL Jr, Hamilton JF & Schmidt EA (1969). Natural landscape preferences: a predictive model. *Journal of Leisure Research 1,* 1–19.
Shanley E (1986). *Mental Handicap: A Handbook of Care.* London: Churchill Livingstone.
Simonton DK (1978). History and the eminent person. *Gifted Child Quarterly 22,* 187–195.
Simonton DK (1989). *Scientific Genius: Psychology of Science.* Cambridge: Cambridge University Press.
Sims ACP, White AC & Murphy T (1979). Aftermath neurosis: psychological sequelae of the Birmingham bombings in victims not seriously injured. *Medicine, Science & the Law 19,* 78–81.
Skeels HM (1966). Adult status of children with contrasting early life experiences: a follow-up study. *Monographs of the Society for Research in Child Development 31, No. 105.*
Skinner BF (1976). *Particulars of my Life.* New York: Knopf.
Sluckin W, Hargreaves DJ & Colman AM (1982). Some experiential studies of familiarity and liking. *Bulletin of the British Psychological Society 35,* 189–194.
Smith D (1989). *North and South.* Harmondsworth: Penguin.
Smith DF & Dorfman DD (1975). The effect of stimulus uncertainty on the relationship between frequency of exposure and liking. *Journal of Personality & Social Psychology 31,* 150–155.
Smith ME (1986). Critique of E Miller "Handedness and a test of cognitive development". *Neuropsychologia 24,* 453–454.
Snyderman M & Rothman S (1988). *The IQ Controversy: The Media and Public Policy.* Oxford: Transaction.
Solzhenitsyn A (1991). *A Day in the Life of Ivan Denisovich.* London: Harvill.
Sommer R & Kroll B (1979). Mental patients and nurses rate habitability. In D Canter & S Canter, *Designing for Therapeutic Environments: A Review of Research.* Chichester: John Wiley.
Spearman CE (1927). *The Abilities of Man.* London: Macmillan.
Spencer H (1876–1896). *Principles of Sociology.* 3 vols. 1969 edition, London: Macmillan.
Spinoza B (1665). *Ethics.*
Spivack M (1967). Sensory distortions in tunnels and corridors. *Hospital & Community Psychiatry 18,* 24–30.
Springer SP & Deutsch G (1989). *Left Brain, Right Brain, 3rd edn.* San Francisco: WH Freeman.
Srole L, Langner TS, Michael ST & Opler MK (1961). *Mental Health in the Metropolis.* New York: McGraw-Hill.
Staats AW (1991). Unified positivism and unification psychology: Fad or new field? *American Psychologist 46,* 899–912.
Stein Z, Susser M, Saenger G & Marolla F (1972). Nutrition and mental performance. *Science 178,* 708–713.
Sternberg RJ (1984). Toward a triarchic theory of human intelligence. *The Behavioral and Brain Sciences 7,* 269–287.
Stott DH (1971). Behavioural aspects of learning disabilities: assessment and remediation. *Experimental Publications System 11,* 400–436.
Surwillo WW (1981). Ear asymmetry in telephone listening behaviour. *Cortex 17,* 625–632.
Sutherland NS (1976). *Breakdown.* London: Weidenfeld & Nicholson.

Sutherland NS (1992). *Irrationality*. London: Constable.
Swan JA (1970). Response to air pollution: a study of attitudes and coping strategies of high school youths. *Environment & Behavior 2,* 127–152.
Swanson JM, Kinsbourne M & Horn JM (1980). Cognitive deficit and left-handedness: a cautionary note. In J Herron, *Neuropsychology of Left-Handedness*. New York: Academic Press.
Szasz T (1973). Mental illness as metaphor. *Nature 242,* 305–307.
Tarnopolsky A & Clark C (1984). Environmental noise and mental health. In H Freeman, *Mental Health and the Environment*. London: Churchill Livingstone.
Teng EL, Lee P-H, Yang K-S & Chang PC (1979). Lateral preferences for hand, foot and eye, and their lack of association with scholastic achievement, in 4143 Chinese. *Neuropsychologia 17,* 41–48.
Terman LM (1925). *Mental and Physical Traits of a thousand Gifted Children. Genetic studies of Genius.* Stanford, CA: Stanford University Press.
Terman LM & Merrill MA (1937). *Measuring Intelligence.* London: Harrap.
Thompson DE, Aiello JR & Epstein EM (1979). Interpersonal distance preference. *Journal of Nonverbal Behaviour 4,* 113–118.
Thomson GH (1951). *The Factorial Analysis of Human Ability, 5th edn.* London: London University Press.
Thoreau HD (1854). *Walden.*
Thorndike RL (1966). Intellectual status and intellectual growth. *Journal of Educational Psychology 57,* 121–127.
Thurstone LL (1924). *The Nature of Intelligence.* London: Kegan Paul, Trench, Trusner.
Tinbergen N (1990). *The Herring Gull's World: A Study of the Social Behaviour of Birds.* London: Collins.
Torrey EF (1990). Offspring of twins with schizophrenia. *Archives of General Psychiatry 47,* 976–977.
Tuan YF (1974). *Topophilia: A Study of Environmental Perception, Attitudes and Values.* Englewood Cliffs, NJ: Prentice-Hall.
Van Liere KD & Dunlap RE (1981). Environmental concern: does it make a difference how it is measured? *Environment & Behavior 13,* 651–676.
Vernon PE (1979). *Intelligence: Heredity and Environment*. San Francisco: WH Freeman.
Von Senden M (1960). *Space and Sight: The Perception of Space and Shape in the Congenitally Blind before and after Operations.* (P Heath trans.) London: Macmillan (orig. work 1932).
Vroon PA (1980). *Intelligence: on Myths and Measurement*. Oxford: North Holland.
Warrington EK, James M & Maciejewski C (1986). The WAIS as a lateralizing and localizing diagnostic instrument: a study of 656 patients with unilateral cerebral lesions. *Neuropsychologia 24,* 223–239.
Watson JD (1980). *The Double Helix*. London: WW Norton.
Weber M (1946). *Essays in Sociology*. Oxford: Oxford University Press.
Wechsler D (1958). *The Measurement and Appraisal of Adult Intelligence.* London: Bailliere, Tindall & Cox.
Weigel R & Weigel J (1978). Environmental concern: the development of a measure. *Environment & Behavior 10,* 3–16.
Wellman JD & Buhyoff GJ (1980). Effects of regional familiarity on landscape preferences. *Journal of Environmental Management 11,* 105–110.

Westfall RS (1981). *Never at Rest: A Biography of Isaac Newton*. Cambridge: Cambridge University Press.
Williams R (1983). *Keywords: the Vocabulary of Culture and Society*. London: Fontana.
Williams S (1982). Dichotic lateral asymmetry: the effects of grammatical structure and telephone usage. *Neuropsychologia 20,* 457–464.
Williams SM (1985). How the familiarity of a landscape affects appreciation of it. *Journal of Environmental Management 21,* 63–67.
Williams SM (1987a). Repeated exposure and the attractiveness of synthetic speech: an inverted-U relationship. *Current Psychological Research & Reviews 6,* 148–154.
Williams SM (1987b). Repeated exposure to computer graphics images: the disappearance of a relationship. *Social & Behavioural Sciences Documents 16,* 52, MS No 7220.
Williams SM (1987c). Differences in academic performance at school depending on handedness: matter for neuropathology? *Journal of Genetic Psychology 148,* 469–478.
Williams SM (1988). *Psychology on the Couch: The Discipline Observed*. Brighton: Harvester.
Williams SM & McCrorie R (1990). Ecological attitudes in town and country. *Journal of Environmental Management 31,* 157–162.
Wilson EO (1975). *Sociobiology: the New Synthesis*. Cambridge, MA: Harvard University Press.
Wilson WR (1979). Feeling more than we can know: exposure effects without learning. *Journal of Personality & Social Psychology 37,* 811–821.
Wing JK (1978). *Reasoning about Madness*. Oxford: Oxford University Press.
Winsborough HH (1965). Social consequences of high population density. *Law and Contemporary Problems 30,* 120–126.
Wittenborn JR (1946). Correlates of handedness among college freshmen. *Journal of Educational Psychology 37,* 161–170.
Wolfensberger W (1972). *Normalization: The Principle of Normalization in Human Services*. Toronto: National Institute of Mental Retardation.
Woodworth RS & Schlosberg H (1955). *Experimental Psychology*. New York: Henry Holt.
Woody E & Claridge G (1977). Psychoticism and creativity. *British Journal of Social & Clinical Psychology 16,* 241–248.
World Health Organisation, Expert Committee on Mental Health. (1953) *Third Report, No 73*. Geneva: World Health Organisation.
Wundt W (1912). *An Introduction to Psychology, 2nd edn*. Translated by R Pintner. New York: Macmillan.
Wundt W (1916). *Völkerpsychologie. Elements of Folk Psychology*. Translated by EL Schaub. London: George Allen & Unwin.
Young RM (1970). *Mind, Brain and Adaptation in the Nineteenth Century: Cerebral Localization and its Biological Context from Gall to Ferrier*. Oxford: Clarendon.
Zajonc RB (1968). Attitudinal effects of mere exposure. *Journal of Personality & Social Psychology Monographs Supplement 9* (No 2, Part 2).
Zajonc RB, Shaver P, Tavris C & Kreveld DV (1972). Exposure, satiation and stimulus discriminability. *Journal of Personality & Social Psychology 21,* 270–280.
Zigler E & Hodapp RM (1986). *Understanding Mental Retardation*. Cambridge: Cambridge University Press.
Ziman J (1976). *The Force of Knowledge*. Cambridge: Cambridge University Press.

Zube EH (1974). Cross-disciplinary and intermode agreement on the description and evaluation of landscape resources. *Environment & Behavior 6*, 69–89.
Zuckerman H (1977). *Scientific Elite: Nobel Laureates in the United States*. New York: Free Press.
Zuckerman H (1978). Theory choice and problem choice in science. In J Gaston, *Sociology of Science: Problems, Approaches and Methods*. London: Jossey-Bass.

Index